Synt
Synt

Standpunkte/Positions

Synthetische Biologie
Synthetic Biology

Stellungnahme/Statement

Standpunkte/Positions

Deutsche Forschungsgemeinschaft
German Research Foundation
Kennedyallee 40 · 53175 Bonn, Germany
Postal address: 53170 Bonn, Germany
Phone: +49 228 885-1
Fax: +49 228 885-2777
postmaster@dfg.de
www.dfg.de

acatech – Deutsche Akademie der Technikwissenschaften
German Academy of Science and Engineering
acatech Head Office: Residenz München, Hofgartenstraße 2, 80539 München, Germany
Phone: +49 89 520309-0, Fax: +49 89 520309-9
info@acatech.de
www.acatech.de

Deutsche Akademie der Naturforscher Leopoldina – Nationale Akademie der Wissenschaften
German Academy of Sciences Leopoldina – National Academy of Sciences
Leopoldina Head Office: Emil-Abderhalden-Straße 37, 06108 Halle/Saale, Germany
Postal address: Postfach 110543, 06019 Halle/Saale, Germany
Phone: +49 345 47239-0, Fax: +49 345 47239-19
leopoldina@leopoldina-halle.de
www.leopoldina-halle.de

Bibliografische Information der Deutschen Nationalbibliothek
Die Deutsche Nationalbibliothek verzeichnet diese Publikation in der Deutschen Nationalbibliografie; detaillierte bibliografische Daten sind im Internet über <http://dnb.d-nb.de> abrufbar.

ISBN 978-3-527-32791-1
© 2009 WILEY-VCH Verlag GmbH & Co. KGaA, Weinheim

Layout und Typografie: Tim Wübben, DFG
Satz: besscom, Berlin
Druck: betz-druck GmbH, Darmstadt
Bindung: Litges & Dopf GmbH, Heppenheim

Printed in the Federal Republic of Germany

Inhalt

Vorwort

Auf Grundlage der Disziplinen Biologie, Molekularbiologie, Chemie, Biotechnologie sowie der Informationstechnologie und Ingenieurwissenschaften entwickelt sich derzeit ein neues Forschungsfeld, das als Synthetische Biologie bezeichnet wird. In jüngster Zeit hat es – auch international – besondere Aufmerksamkeit erlangt.

Die Synthetische Biologie kann wesentlich zum Erkenntnisgewinn in der Grundlagenforschung beitragen. Darüber hinaus eröffnet sie mittelfristig Möglichkeiten biotechnologischer Anwendungen, wie zum Beispiel im Bereich neuer und verbesserter Diagnostika, Impfstoffe und Medikamente oder auch bei der Entwicklung neuer Biosensoren oder Biomaterialien bis hin zu Biokraftstoffen.

Gleichzeitig wirft das Forschungsgebiet neue Fragen auf, zum Beispiel zu rechtlichen Aspekten im Rahmen der biologischen Sicherheit oder dem Schutz vor Missbrauch, ebenso zur wirtschaftlichen Verwertung und zu ethischen Aspekten.

Vor diesem Hintergrund haben die Deutsche Forschungsgemeinschaft (DFG), acatech – Deutsche Akademie der Technikwissenschaften – und die Deutsche Akademie der Naturforscher Leopoldina – Nationale Akademie der Wissenschaften – ihre Kräfte gebündelt und eine gemeinsame Stellungnahme zu den möglichen Chancen und Risiken der Synthetischen Biologie erarbeitet.

Um einen konstruktiven Dialog zwischen den Disziplinen anzuregen, wurde von den drei Organisationen ein gemeinsamer, internationaler Workshop initiiert. Wissenschaftlerinnen und Wissenschaftler aus den Bereichen Biochemie, Molekularbiologie, Genetik, Mikrobiologie, Virologie, der Chemie und Physik sowie aus den Sozial- und Geisteswissenschaften trafen sich zu einem Informationsaustausch, ergänzt durch Vertreterinnen und Vertreter aus öffentlichen Einrichtungen und der Industrie. Die Informationen aus den Vorträgen und den ausführlichen Diskussionsrunden bilden die Grundlage für die folgende Stellungnahme. Diese richtet sich an Vertreterinnen und Vertreter der Politik und Behörden, an die Öffentlichkeit und nicht zuletzt an die wissenschaftliche Gemeinschaft.

Die Synthetische Biologie konzentriert sich derzeit noch überwiegend auf die Grundlagenforschung.

Wie bei jeder neuen Technologie, die einen bedeutenden Einfluss entwickelt, ist neben den wirtschaftlichen Chancen und dem wissenschaftlichen Forschungsinteresse auch die Frage der nicht beabsichtigten Nebenfolgen frühzeitig zu behandeln. Dies bedeutet vor allem, dass Risiken und Chancen, soweit möglich, abgeschätzt werden und die Lehren daraus bereits in das Design und die Anwendungsbedingungen der neuen Technologie einfließen müssen. Zudem ist der frühzeitige und offene Dialog mit der Öffentlichkeit wie bei jeder neuen Technologie wichtig. Nur so kann in einer demokratischen und pluralen Gesellschaft ein verantwortungsvolles Innovationsklima geschaffen werden.

So ist nicht nur die Hoffnung auf Erkenntnis groß, sondern auch der Bedarf für eine breite wissenschaftliche sowie öffentliche Erörterung der Fragen bei zukünftigen Anwendungsmöglichkeiten, da die Chancen und Herausforderungen einer sorgfältigen Abwägung unterzogen werden sollen.

Juli 2009

Prof. Dr.-Ing.
Matthias Kleiner

Präsident
Deutsche
Forschungsgemeinschaft

Prof. Dr.
Reinhard Hüttl

Präsident
acatech –
Deutsche Akademie
der Technikwissenschaften

Prof. Dr.
Volker ter Meulen

Präsident
Deutsche Akademie der
Naturforscher Leopoldina –
Nationale Akademie
der Wissenschaften

1   Zusammenfassung und Empfehlungen

Die Synthetische Biologie basiert auf den Erkenntnissen der molekularen Biologie, der Entschlüsselung kompletter Genome, der ganzheitlichen Betrachtung biologischer Systeme und dem technologischen Fortschritt bei der Synthese und Analyse von Nukleinsäuren. Sie führt ein weites Spektrum an naturwissenschaftlichen Disziplinen zusammen und verfolgt dabei ingenieurwissenschaftliche Prinzipien. Das spezifische Merkmal der Synthetischen Biologie ist, dass sie biologische Systeme wesentlich verändert und gegebenenfalls mit chemisch synthetisierten Komponenten zu neuen Einheiten kombiniert. Dabei können Eigenschaften entstehen, wie sie in natürlich vorkommenden Organismen bisher nicht bekannt sind.

Die Synthetische Biologie steht für ein Forschungs- und Anwendungsgebiet, das sich nicht strikt von den herkömmlichen gentechnischen und biotechnologischen Verfahren unterscheidet und deshalb als eine Weiterentwicklung dieser Disziplinen und der damit verfolgten Ziele verstanden werden kann. Die vorliegende Stellungnahme behandelt im ersten Teil ausgewählte grundlagenorientierte Gebiete der Synthetischen Biologie:

▶ Die technologischen Forschritte bei der Synthese und Analyse von Nukleinsäuren. Durch sie werden nicht nur die Verfahren der rekombinanten Gentechnik erleichtert, sondern auch erhebliche Fortschritte bei der Gentherapie eröffnet.

▶ Die Konstruktion von Minimalzellen mit synthetisch hergestellten oder genetisch verkleinerten Genomen mit dem Ziel, eine kleinste lebensfähige Einheit zu gewinnen. Derartige Zellen sind unter definierten Laborbedingungen lebensfähig, haben jedoch eingeschränkte Fähigkeiten, sich an natürlichen Standorten zu vermehren.

▶ Die Synthese von Protozellen mit Merkmalen lebender Zellen. Es ist beabsichtigt, sie langfristig – ebenso wie die Minimalzellen – als „Chassis" für die Herstellung von Substanzen einzusetzen.

▶ Die Produktion neuer Biomoleküle durch baukastenartiges Zusammenfügen einzelner Stoffwechselfunktionen. Diese können aus verschiedensten genetischen Spenderorganismen stammen.

▶ Die Konstruktion regulatorischer Schaltkreise, die auf externe Reize reagieren. Diese erlauben es, komplexe biologische oder synthetische Prozesse zu steuern.

▶ Die Konzeption sogenannter „orthogonaler Systeme". Dabei werden modifizierte Zellmaschinerien eingesetzt, um beispielsweise neuartige Biopolymere zu erzeugen.

Die gegenwärtigen Arbeiten auf dem Gebiet der Synthetischen Biologie bewegen sich überwiegend noch auf der Ebene der Grundlagenforschung. Es ist zu erwarten, dass daraus wichtige wissenschaftliche Erkenntnisse resultieren werden, die die Entwicklung von neuen Medikamenten und Therapieverfahren sowie die Produktion von Industriechemikalien und die Konzeption von katalytischen Prozessen nachhaltig beeinflussen. Damit kann es der Synthetischen Biologie gelingen, Organismen herzustellen, die nur unter kontrollierten Bedingungen überleben können.

Wie ist das Marktpotenzial einzuschätzen? Welches sind die wissenschaftlichen Rahmenbedingungen? Birgt die Synthetische Biologie neben diesen vielfältigen Chancen auch mögliche Risiken? Diese Fragen werden aus aktueller Sicht im zweiten Teil der Stellungnahme behandelt. Dabei werden folgende Aspekte diskutiert:

▶ Die ökonomische Bedeutung der Synthetischen Biologie lässt sich derzeit zwar noch nicht präzise abschätzen; es sind jedoch bereits marktnahe Produkte erkennbar, die sowohl für die industrielle Verwertung als auch den gesellschaftlichen Nutzen vielversprechende Perspektiven bieten. Der Katalog umfasst Medikamente, Nukleinsäure-Vakzine, neuartige Verfahren zur Gentherapie, umwelt- und ressourcenschonende Fein- und Industriechemikalien, Biobrennstoffe sowie neue Werkstoffe wie polymere Verbindungen.

▶ Die wissenschaftlichen Rahmenbedingungen für die Synthetische Biologie in Deutschland werden als günstig eingeschätzt. Es gibt sowohl auf europäischer als auch nationaler Ebene erste Förderprogramme, die diese Disziplinen gezielt berücksichtigen. Durch die Überlappung mit konventionell biotechnologischen und molekularbiologischen Vorhaben werden Projekte der Synthetischen Biologie auch in anderen Themenschwerpunkten gefördert. Grundlegende Infrastrukturen sind vorhanden oder in existierenden Forschungszentren ausbaufähig. Eine positive Ausgangssituation wird in der Stärke der Fachrichtungen Chemie und Mikrobiologie gesehen. Die interdisziplinäre Ausrichtung der Synthetischen Biologie erfordert ein abgestimmtes Ausbildungskonzept für Naturwissenschaftlerinnen und Naturwissenschaftler sowie Ingenieurinnen und Ingenieure.

▶ Ähnlich wie bei der Gentechnik, aber auch der konventionellen Züchtung treten bei der Synthetischen Biologie Risiken in Bezug auf biologische Sicherheit (Biosafety) oder in Bezug auf Missbrauchsmöglichkeiten (Biosecurity) auf. Es ist noch eine offene Frage, ob die Risiken der Synthetischen Biologie anders gelagert oder in ihrer Größenordnung anders einzuschätzen sind als die Risiken der bisherigen Genforschung. Zunächst ist davon auszugehen, dass die bestehenden Regelungen und Regulierungen ausreichen, um diese Risiken zu vermeiden oder abzumildern. Wichtig ist aber eine gesellschaftliche Begleitforschung, die frühzeitig neue Risiken erkennen hilft, damit mögliche Fehlentwicklungen von vornherein vermieden werden können. In Bezug auf die biologische Sicherheit sind die Risiken der gegenwärtigen Forschung innerhalb der Synthetischen Biologie durch gesetzliche Regelungen angemessen erfasst und reguliert. Einige der in der Synthetischen Biologie verwendeten Ansätze tragen sogar zu einer Erhöhung der biologischen Sicherheit im Umgang mit genetisch modifizierten Organismen bei. Ein mögliches Missbrauchspotenzial der Synthetischen Biologie stellt der kommerzielle Erwerb von DNA-Sequenzen dar, basierend auf öffentlich verfügbaren Genomdaten. In Deutschland existieren aber schon heute gesetzliche Regelungen, die dieses Missbrauchsrisiko einschränken (Gentechnikgesetz, Infektionsschutzgesetz, Kriegswaffenkontrollgesetz, Außenwirtschaftsgesetz). Neben den gesetzlichen Regelungen existieren noch freiwillige Selbstverpflichtungen, die innerhalb der wissenschaftlichen Gemeinschaft und der Industrie beim Umgang mit Toxinen und Krankheitserregern sowie bei der Überprüfung der Seriosität der Besteller von Nukleinsäuresequenzen gelten. Auch haben sich Forscher und Hersteller synthetischer Nukleinsäuren darauf verständigt, die potenziellen Gefahren, die von den angeforderten Nukleinsäurepräparaten ausgehen könnten, zu bestimmen und durch geeignete Maßnahmen zu entschärfen.

▶ Weil sich bei einigen Anwendungen die Grenzen zwischen Lebendigem und Technisch-Konstruiertem verwischen, hat dies in der Öffentlichkeit zu der Besorgnis geführt, dass hier der Mensch ethische Grenzen überschreite. Dabei wird argumentiert, dass die Identität des Lebendigen leide, wenn neuartiges Leben geschaffen werde, und dass sich der Mensch durch solche Eingriffe zum Schöpfer aufspiele. Dem wird entgegengehalten, dass eine Beeinflussung der natürlichen Evolution keineswegs grundsätzlich ethisch unzulässig sei und auch nicht den Respekt vor dem Leben schmälern müsse. Mit der Anwendung der Synthetischen Biologie sind zudem erhebliche Nutzenpotenziale verbunden, wie etwa für die Medizin oder den Umweltschutz. Aus ethischer Sicht bedarf es einer angemessenen Beurteilung und Abwägung gegen mögliche Risiken der Synthetischen Biologie. Solche und andere Fragen müssen im Diskurs mit allen gesellschaftlichen Gruppen erörtert werden.

Als Resümee der vorliegenden Stellungnahme werden folgende Empfehlungen gegeben:

(1) Die Synthetische Biologie stellt eine konsequente Weiterentwicklung bestehender Methoden der molekularen Biologie dar und besitzt ein großes Innovationspotenzial, von dem sowohl die Grundlagenforschung als auch die industrielle Anwendung profitieren werden. Da sich die anwendungsbezogenen Projekte vorwiegend noch auf konzeptionellen Ebenen bewegen, sollte die Grundlagenforschung gestärkt und zukünftig bei der Planung wissenschaftlicher Förderprogramme Berücksichtigung finden.

(2) Der Erfolg der Synthetischen Biologie wird maßgeblich davon abhängen, inwieweit es gelingen wird, die verschiedensten Disziplinen in Forschungszentren und Forschungsverbünden zusammenzuführen und Infrastrukturen zu bündeln. Darüber hinaus sollten angehende junge Wissenschaftlerinnen und Wissenschaftler im Rahmen des Bachelor-, Master- und Graduiertenstudiums mit der Thematik vertraut gemacht und durch Öffnung neuer beruflicher Perspektiven auf das Fachgebiet vorbereitet werden.

(3) Bei der ökonomischen Verwertung der Synthetischen Biologie ist zu beachten, dass diese nicht nur von einer starken, im internationalen Wettbewerb konkurrenzfähigen Forschung abhängt, sondern dass auch die rechtlichen und die gesellschaftlichen Rahmenbedingungen mitbestimmend für den Erfolg oder

Misserfolg dieser neuen Technologie sind. Für eine wirtschaftlich erfolgreiche Verwertung der neuen Technologie sowie für ihre gesellschaftliche Akzeptanz ist eine frühzeitige Begleitforschung zu den Chancen und Risiken sinnvoll. Dabei gilt es, das technische Design sozialverträglich auszuloten, um eine Verstärkung der Chancen und eine Minderung der Risiken herbeizuführen. Die wirtschaftliche Verwertung der im Rahmen der Synthetischen Biologie entwickelten Verfahren und Produkte sollte prinzipiell dem gleichen patentrechtlichen Schutz unterliegen, der auch für die herkömmlichen rekombinanten Genprodukte oder Genfragmente gilt. Minimalzellen und Protozellen sollten urheberrechtlich geschützt werden können (am besten durch Patente), um einen wirtschaftlichen Anreiz für Investitionen in neue Techniken zu geben.

(4) Bezüglich der biologischen Sicherheit (Biosafety) und des Missbrauchsrisikos (Biosecurity) sind die bestehenden Gesetze in Deutschland nach dem heutigen Forschungsstand ausreichend. Aufgrund der dynamischen und vielfältigen Entwicklungen wird jedoch empfohlen,

▶ die Zentrale Kommission für die Biologische Sicherheit (ZKBS) zu beauftragen, ein wissenschaftliches Monitoring durchzuführen, um die aktuellen Entwicklungen sachverständig und kritisch zu begleiten und

▶ für Freisetzungen und Handhabung in geschlossenen Systemen von Organismen der Synthetischen Biologie, die keinen Referenzorganismus in der Natur haben, klar definierte Kriterien zur Risikoabschätzung festzulegen.

Zur Reduzierung des Missbrauchsrisikos wird vorgeschlagen,

▶ eine Kontaktstelle mit einer standardisierten Datenbank zur Überprüfung der DNA-Sequenzen einzurichten, an die sich Unternehmen bei fragwürdigen Bestellungen wenden können und

▶ Mitarbeiterinnen und Mitarbeiter im Rahmen von Unterweisungen nach der Gentechnik-Sicherheitsverordnung (GenTSV) über mögliche Missbrauchsrisiken der Synthetischen Biologie aufzuklären.

▶ Sollten sich zusätzliche Regeln für die Risikobewertung, Überwachung und Kontrolle der Forschung und Anwendung der Synthetischen Biologie im Verlauf der Entwicklung als notwendig herausstellen, so wird empfohlen, diese in Form von international anerkannten Grundsätzen zu verfassen, die Vorbild für nationale Regelungen sein könnten.

(5) Soweit bewährte Methoden der Technikfolgenbeurteilung und der Risikoanalyse nicht greifen oder bei den zu erwartenden Auswirkungen hohe Unsicherheiten herrschen, muss das Vorsorgeprinzip gelten. Außerdem ist es ratsam, durch die Schaffung geeigneter interdisziplinärer Diskussionsplattformen die Selbstkontrolle der Wissenschaft zu fördern. Für Fragen der ethischen Beurteilung von technisch konstruierten Lebensformen sollte möglichst zeitnah ein öffentlicher Dialog geführt werden. In diesem Dialog sollten die Argumente ausgetauscht und die verschiedenen Interpretationen des Lebendigen gegenüber dem Nichtlebendigen diskutiert werden. Als Ziel des Diskurses ist die ethische Bewertung kopierender oder auch de novo synthetisierender Interventionen in die vorgefundene Natur anzustreben.

2 Einführung

In einem interdisziplinären Umfeld von Biologie, Chemie, Physik, Mathematik, Ingenieurwissenschaften, Biotechnologie und Informationstechnik verstärkt sich seit wenigen Jahren eine Forschungsrichtung, die als Synthetische Biologie bezeichnet wird.[1,2,3,4,5] Wissenschaftler der unterschiedlichsten Fachrichtungen arbeiten dabei zusammen, um biologische Systeme mit neuen, definierten Eigenschaften zu konzipieren. Dabei sollen die Systeme vornehmlich künstlich hergestellt bzw. nachgebaut werden, mit dem Ziel, neue biologische Komponenten sowie neuartige lebende Organismen, die in der Natur in dieser Form nicht bekannt sind, zu gewinnen. Geleitet von ingenieurwissenschaftlichen Prinzipien, werden dabei fortgeschrittene Methoden der Molekularbiologie, der rekombinanten Gentechnik und der chemischen Synthese von biologischen Bausteinen vereint. Basierend auf einem von Menschen entworfenen rationalen Design sollen durch die Zusammenführung von synthetischen und biologischen Einheiten neue Stoffe und Systeme, zum Beispiel neuartige polymere Moleküle, Gewebe, ganze Zellen und Organismen, geschaffen werden.

Sind diese der Synthetischen Biologie zugrunde liegenden Strategien und die daraus resultierenden Produkte tatsächlich revolutionär neu? Bereits im Jahr 1912 erschien in der Veröffentlichung von Stéphane Leduc der Begriff „La Biologie Synthétique"[6] und im gleichen Jahr formulierte Jacques Loeb, dass es möglich sein sollte, künstliche lebende Systeme zu generieren.[7] Nach der Verwendung des Begriffs „Synthetische Biologie" in den Ausführungen von Waclaw Szybalski[8] wird der heutige Sinninhalt der Synthetischen Biologie vor allem geprägt durch den Bericht von Eric Kool aus dem Jahr 2000 zum

---

1   Hartwell LH, Hopfield JJ, Leibler S, Murray AW; From molecular to modular cell biology. Nature, 1999, 402, C47–C52.

2   Benner SA, Sismour AM; Synthetic Biology. Nat. Rev. Genet., 2005, 6, 533–543.

3   Endy D; Foundations for engineering biology. Nature, 2005, 438, 449–453.

4   Andrianantoandro E, Basu S, Karig DK, Weiss R; Synthetic biology: new engineering rules for an emerging discipline. Mol. Syst. Biol., 2006, 2, 0028.

5   Heinemann M, Panke S; Synthetic Biology – putting engineering into biology. Bioinformatics, 2006, 22, 2790–2799.

6   Leduc S; La biologie synthétique. In: Études de biophysique. A. Poinat (ed.), Paris, 1912.

7   Loeb J; The mechanistic conception of life. In: Biological Essays. University of Chicago Press, Chicago, 1912.

8   Szybalski W; In vivo and in vitro Initiation of Transcription, 405. In: A. Kohn and A. Shatkay (Eds.), Control of Gene Expression, 23–24, and Discussion, 404–405 (Szybalski's concept of Synthetic Biology), 411–412, 415–417. New York: Plenum Press, 1974.

Einbau von künstlichen chemischen Komponenten in biologische Systeme.[9] Durch die technologischen Innovationen bei Nukleinsäuresynthesen und DNA-Sequenzierungen hat das Gebiet der Synthetischen Biologie zweifelsfrei einen rasanten dynamischen Verlauf genommen. Es besteht allerdings ein fließender Übergang zwischen der Synthetischen Biologie und den seit über 30 Jahren im Einsatz befindlichen gentechnologischen Verfahren, zum Beispiel zur Gewinnung von rekombinanten Genprodukten.

Das Potenzial der Synthetischen Biologie ist weit gefächert. Die Forschungsrichtung trägt erheblich zum Erkenntnisgewinn auf der Ebene der Grundlagenforschung bei, indem sie zum Beispiel versucht, Antworten auf die Frage nach den Voraussetzungen für die Lebensfähigkeit von Zellen zu liefern. Darüber hinaus eröffnet die Synthetische Biologie neue Möglichkeiten biotechnologischer Anwendungen, beispielsweise die Entwicklung verbesserter, auf den individuellen Patienten zugeschnittener Pharmaka, Impfstoffe und Diagnostika, die Bereitstellung synthetischer Genvektoren für eine erfolgreiche Gentherapie sowie die Konzeption spezifischer Biosensoren, biologischer Brennstoffzellen und Zellfabriken für die Produktion neuartiger Biomaterialien. Die Synthetische Biologie umfasst Verfahren zur großtechnischen Gewinnung von Biobrennstoffen wie Ethanol, Methanol und Wasserstoff und zur Beseitigung umweltschädlicher Substanzen. Sie strebt an, Organismen in ihren Merkmalen so gezielt zu verändern, dass sie, mit grundlegend neuen, vom Menschen entworfenen Eigenschaften versehen, besondere Leistungen vollbringen.

Die vorliegende Stellungnahme zielt in Kapitel 3 darauf ab, zunächst den naturwissenschaftlichen Hintergrund für ausgewählte Bereiche der Synthetischen Biologie zu vermitteln und die Bedeutung für den allgemeinen wissenschaftlichen Erkenntnisgewinn aufzuzeigen. Sechs Themenkomplexe werden vertiefend diskutiert:

► Die chemisch-enzymatische Synthese von Nukleinsäuren bis hin zu kompletten Genomen. Sie ist ein Instrumentarium, mit dem Gensequenzen gezielt optimiert und verändert werden können. Entstehende Produkte können beispielsweise bei der Herstellung von DNA-Vakzinen und in der somatischen Gentherapie zum Einsatz kommen.

► Die Konstruktion von Zellen mit einem Minimalgenom. Diese auch als „Chassis" bezeichnete genetische Plattform trägt das Mindestmaß an unentbehrlichen Informationen für die Lebensfähigkeit einer Zelle. Minimalzellen geben Aufschluss über die evolutionäre Anpassung von Organismen an natürliche Standorte.

► Die Synthese von Protozellen. Deren Bauplan folgt entweder biologischen oder physikalischen Prinzipien. Protozellen können als Modelle lebender Zellen betrachtet werden.

► Die Produktion von Biomolekülen in einem bisher noch nicht verfügbaren Maßstab. Durch gentechnisches Zusammenfügen kompletter Stoffwechsel-

---

9   Vgl. Rawls R; Synthetic Biology makes its debut. Chem. Eng. News, 2000, 78, 49–53.

reaktionswege nach dem Baukastenprinzip („BioBricks") kann es gelingen, neuartige Substanzen oder Produktionsformen zu entwickeln.

▶ Die Konzeption von regulatorischen Schaltkreisen („regulatory circuits"). Sie sind mit empfindlichen sensorischen Funktionen ausgestattet und können netzartig zelluläre oder industrielle Prozesse steuern.

▶ Der Einsatz modifizierter zellulärer Maschinen im Rahmen der sogenannten „orthogonalen Systeme". Diese Vorgehensweise erlaubt beispielsweise die Herstellung von polymeren Verbindungen aus chemischen Bausteinen nach dem Reißbrettprinzip.

Das umfangreiche, zum Teil noch visionäre Spektrum der Synthetischen Biologie wirft zudem eine Vielzahl von Fragen auf, die in Kapitel 4 der Stellungnahme angesprochen werden:

▶ Worin besteht der wirtschaftliche Nutzen der Synthetischen Biologie und inwieweit profitiert die Gesellschaft von den neuen Entwicklungen?

▶ Besteht die Gefahr der Entstehung von Monopolen auf diesem Forschungsgebiet?

▶ Geht von der Synthetischen Biologie ein besonderes Risikopotenzial aus, das zusätzliche Sicherheitsvorkehrungen erfordert, oder reichen die vorhandenen gesetzlichen Bestimmungen und die dafür zuständigen Überwachungsgremien für den Einsatz der Synthetischen Biologie aus?

▶ Welche ethischen Überlegungen begleiten die Synthetische Biologie, insbesondere solche Projekte, die auf die Herstellung synthetischer Zellen abzielen oder die Freisetzung neuartiger Organismen vorsehen?

3 Ausgewählte Forschungsfelder

3.1 Chemische Synthesen von Genen und Genomen

Zu den wichtigen Fortschritten in Richtung einer Synthetischen Biologie gehört, dass DNA beliebiger Sequenz und fast beliebiger Länge ohne Matrize synthetisiert werden kann und damit die *de-novo*-Synthese von Genen und sogar ganzer Genome möglich geworden ist. Damit lassen sich neue biologische Funktionen prinzipiell auf dem Reißbrett entwerfen und für Forschungs- und Anwendungszwecke einsetzen. Eine unschätzbare Hilfe dabei sind die Informationen, die durch die neuen Hochdurchsatz-Sequenziertechnologien bereitgestellt werden.[10,11]

In der herkömmlichen Oligonukleotidsynthese werden kurzkettige Einzelstrang-DNA-Moleküle (~5 bis ~50 Nukleotide) in automatisierten Prozessen sequenzspezifisch synthetisiert. Die Gensynthese verknüpft mehrere Oligonukleotide mittels gestaffelter Polymerasekettenreaktionen, Chip-basierter Methoden oder den Zusammenbau an der Festphase und Plasmidklonierungen zu langkettigen synthetischen DNA-Sequenzen.[12] Somit können mehrere Kilobasen (kb) Erbinformation gemäß der Sequenzvorgabe des Experimentators erzeugt werden. Die Maximalvariante der Gensynthese ist die Genomsynthese, bei der die gesamte Erbinformation von Viren oder Bakterien und künftig auch die minimaler eukaryonter Genome (s. Kap. 3.2) neu aufgebaut wird. Spektakuläre Beispiele der letzten Zeit sind die Totalsynthese des Poliomyelitis-(Kinderlähmungs-)Virus-Genoms (~7,5 kb)[13] und des sehr viel größeren *Mycoplasma*-Genoms (~583 kb).[14]

Die neuen Möglichkeiten zur Synthese definierter, großer DNA-Fragmente werden die gesamte lebenswissenschaftliche Forschung entscheidend beeinflussen. Langkettige DNA-Sequenzen werden kommerziell und in hoher Qualität

10   Hall N; Advanced sequencing technologies and their wider impact in microbiology. J. Exp. Biol., 2007, 210, 1518–1525.

11   Church GM; Genomes for all. Sci. Am., 2006, 294, 46–54.

12   Tian J, Gong H, Sheng N, Zhou X, Gulari E, Gao X, Church G; Accurate multiplex gene synthesis from programmable DNA microchips. Nature, 2004, 432, 1050–1054.

13   Cello J, Aniko VP, Wimmer E; Chemical synthesis of poliovirus cDNA: Generation of infectious virus in the absence of natural template. Science, 2002, 297, 1016–1018.

14   Gibson DG, Benders GA, Andrews-Pfannkoch C, Denisova EA, Baden-Tillson H, Zaveri J, Stockwell TB, Brownley A, Thomas DW, Algire MA, Merryman C, Young L, Noskov VN, Glass JI, Venter JC, Hutchison CA 3rd, Smith HO; Complete chemical synthesis, assembly, and cloning of a Mycoplasma genitalium genome. Science, 2008, 319, 1215–1220.

für jedes Labor und für nahezu jede Anwendung zugänglich. Dies wird langfristig zur Einsparung von Finanzmitteln und zur Verminderung des Zeitaufwands für die Herstellung genetischer Konstrukte führen.

Die Möglichkeit, Erbinformation von potenziell hoch pathogenen Viren durch DNA-Synthesen zu erzeugen, birgt aber auch Gefahren von Missbrauch. In diesem Zusammenhang hat die gezielte Bestellung „biowaffengeeigneter" DNA-Sequenzen durch einen britischen Reporter Aufsehen erregt.[15] Daher unterliegen die Anbieter von DNA-Synthesen besonderen Auflagen, deren Inhalte Gegenstand laufender Diskussionen sind. Führende kommerzielle Anbieter von DNA-Synthesen versuchen, potenziellen Gefahren durch selbstverpflichtende Kodices vorzubeugen (s. Kap. 4.3).

Die technologischen Grundlagen der Gensynthese wurden vor mehr als 20 Jahren etabliert. Technologische Fortschritte steigern die Produktivität und Qualität der Prozesse und senken kontinuierlich die Kosten. Heute bieten weltweit mehrere Dutzend Firmen kommerziell DNA-Synthesen an. Darunter befinden sich marktführende Unternehmen in Europa (Deutschland) und den USA. Während kurze DNA-Fragmente von 0,1 bis 1 kb innerhalb weniger Tage lieferbar sind, kann die Synthese eines relativ großen Genoms (zum Beispiel ein hypothetisches Minimalgenom in der Größe von ~110 kb)[16] mit allen notwendigen Qualitätskontrollen derzeit bis zu einem Jahr dauern. Zum Vergleich: Das Genom des Bakteriums *Escherichia coli* K-12 umfasst ~4,6 Mb und das menschliche Genom ~3000 Mb.

Die chemische Synthese von DNA ermöglicht ferner die Entwicklung neuartiger, sequenzoptimierter DNA-Bibliotheken oder den Aufbau rekombinanter Gensequenzen, die mehrere künstlich zusammengefügte funktionelle Domänen vereinigen. So dient die DNA-Synthese der Herstellung Kodon-optimierter Varianten menschlicher cDNAs, die unter Beibehaltung der natürlichen Aminosäuresequenz über bessere Expressionseigenschaften nach dem Gentransfer in menschliche oder nicht menschliche Zellen verfügen.

Die gegenwärtige Anwendung synthetischer DNA im Bereich der Medikamentenentwicklung betrifft DNA-Vakzine und die somatische Gentherapie.

Im ersten Anwendungsbeispiel, DNA-Vakzine, werden diese wie herkömmliche Impfstoffe eingesetzt und führen zur Bildung von Antigenen unter Nutzung der körpereigenen Proteinsynthesemaschinerie. Die so gebildeten Antigene rufen ihrerseits eine Immunantwort hervor. Für den Hersteller der Vakzine entfiele die Produktion und Reinigung der Antigene im großen Maßstab. Dadurch würde eine größere Flexibilität bei der Auswahl antigener Proteine eröffnet. Beispielsweise ist es durch die Synthese einer Kodon-optimierten Variante von Genombereichen des humanen Immundefizienzvirus Typ 1 (HIV-1) gelungen, einen komplexen DNA-Impfstoff gegen HIV-1 zu erzeugen, das mul-

---

15   Randerson J; Revealed: the lax laws that could allow assembly of deadly virus DNA. The Guardian, 14 June 2006; www.guardian.co.uk/world/2006/jun/14/terrorism.topstories3

16   Forster AS, Church GM; Towards synthesis of a minimal cell. Mol. Syst. Biol., 2006, 2, 45.

tiple Antigene präsentieren kann.[17] Deren potenzielle Eignung zur Prävention der HIV-1-Infektion muss noch in umfangreichen klinischen Studien geprüft werden. Für die Herstellung von DNA-Vakzinen im großtechnischen Maßstab auf dem Wege der chemischen Synthese ist allerdings der Kostenfaktor derzeit noch viel zu hoch.

Im zweiten Anwendungsbeispiel, der somatischen Gentherapie, ist beabsichtigt, den Transfer rekombinanter DNA in Körperzellen zu nutzen, um Krankheiten zu lindern oder zu heilen. Zahlreiche Anwendungen befinden sich im Stadium der präklinischen Entwicklung oder klinischen Prüfung für Indikationen wie Krebs und entzündliche, degenerative oder monogene Erkrankungen. Zudem werden mittels der Synthese langer Genabschnitte auch neuartige, *in silico* konzipierte Aminosäuresequenzen leichter zugänglich. Solche „Designerproteine" können beispielsweise antivirale Aktivität aufweisen. Wie bei allen Anwendungen der somatischen Gentherapie müssen die biologischen Eigenschaften und möglichen toxikologischen oder immunologischen Reaktionen in umfangreichen präklinischen Studien evaluiert werden, bevor der Einsatz am Menschen möglich wird.

3.2 Entwicklung von Minimalzellen – Zellen reduziert auf essenzielle Lebensfunktionen

Die Synthetische Biologie verfolgt als eines ihrer Ziele die Entwicklung von sogenannten Minimalzellen, die nur unbedingt lebensnotwendige Komponenten enthalten. Minimalzellen sind durch ihre Minimalgenome definiert. Ein Minimalgenom enthält nur solche Gene, die für ein Leben eines bestimmten Organismus unter definierten Bedingungen benötigt werden. Durch die Generierung von Minimalzellen wird zum einen ausgelotet, unter welchen Bedingungen welche Gene einer lebenden Zelle essenziell sind, und zum anderen eine Plattform („Chassis") für den Aufbau neuer Funktionen geschaffen.

Umfangreiche Genomsequenzierungsprojekte haben in der Zwischenzeit gezeigt, dass bakterielle Genome stark in ihrer Größe variieren. Die ersten bakteriellen Genome, die 1995 in den USA sequenziert wurden, betreffen das *Haemophilus-influenzae*-Genom mit 1,83 Mb[18] und das *Mycoplasma-genitalium*-Genom mit 0,58 Mb[19]. Mit der Sequenzierung des 0,82 Mb großen *Mycoplasma-*

17   Bojak A, Wild J, Deml L, Wagner R. Impact of codon usage modification on T cell immunogenicity and longevity of HIV-1 gag-specific DNA vaccines. Intervirology, 2002, 45, 275–286.

18   Fleischmann RD, Adams MD, White O, Clayton, RA, Kirkness, EF, Kerlavage AR, Bult CJ, Tomb JF, Dougherty BA, Merrick JM et al.; Whole-genome random sequencing and assembly of Haemophilus influenzae Rd. Science, 1995, 269, 496–512.

19   Fraser CM, Gocayne JD, White O, Adams MD, Clayton RA, Fleischmann RD, Bult CJ, Kerlavage AR, Sutton G, Kelley JM, Fritchman RD, Weidman JF, Small KV, Sandusky M, Fuhrmann J, Nguyen D, Utterback TR, Saudek DM, Phillips CA, Merrick JM, Tomb JF, Dougherty BA, Bott KF, Hu PC, Lucier TS, Peterson SN, Smith HO, Hutchison CA 3rd, Venter JC; The minimal gene complement of Mycoplasma genitalium. Science, 1995, 270, 397–403.

pneumoniae-Genoms[20] zählt auch eine deutsche Gruppe zu den Pionieren der bakteriellen Genomforschung. In der Zwischenzeit wurden noch deutlich kleinere bakterielle Genome sequenziert: Das *Nanoarchaeum-equitans*-Genom[21] hat eine Größe von 0,49 Mb und das *Buchnera-aphidicola*-Genom[22] von 0,42 Mb. Als kleinstes bakterielles Genom wird heute das Genom des Endosymbionten *Carsonella ruddii*[23] gehandelt, das nur noch ~0,16 Mb misst. Für alle diese Bakterien gilt, dass ihr an bestimmte Wirte angepasster Lebensstil ihre geringe Genomgröße bedingt. Allerdings zieht dieser Lebensstil auch nach sich, dass diese Bakterien experimentell schwierig zu handhaben sind, was einen großen Nachteil bei der Aufklärung essenzieller Lebensfunktionen darstellt.

Zur Entwicklung von Minimalgenomen kann ein *top-down*- oder ein *bottom-up*-Ansatz gewählt werden. Der *top-down*-Ansatz nutzt die gezielte Reduktion vorhandener Genome, während der *bottom-up*-Ansatz das Minimalgenom aus einzelnen DNA-Fragmenten aufbaut.

Mit der Erzeugung von Minimalzellen verfolgt die Synthetische Biologie zunächst ein wissenschaftliches Ziel. Es sollen vereinfachte zelluläre Systeme generiert werden, die es erleichtern, über die parallele Erfassung von Transkriptom-, Proteom- und Metabolom-Daten das systematische Zusammenspiel von essenziellen Zellmodulen mithilfe der mathematischen Modellierung im Rahmen der Systembiologie zu verstehen.

Zusätzlich ist noch ein anwendungsorientiertes Ziel von Interesse, das die Verwendung von Minimalzellen für unterschiedliche biotechnologische Produktionsprozesse vorsieht. In das Minimalgenom einer Zelle, die als „Chassis" genutzt wird, können genetische Komponenten für gewünschte Stoffwechselleistungen eingebaut und im Hinblick auf eine effiziente Produktion optimiert werden. Im Weiteren spielt bei der beschriebenen Entwicklung von Produktionsstämmen auch der biologische Sicherheitsaspekt eine Rolle. Zunächst wird auf dem Weg zur Erzeugung von Minimalgenomen darauf zu achten sein, dass diese Minimalgenome keine Pathogenitätsdeterminanten tragen. Darüber hinaus ist von großer Bedeutung, dass die Vermehrungsfähigkeit von Minimalzellen in der natürlichen Umwelt stark reduziert ist, da dem Minimalgenom ja gerade all die Gene fehlen, die eine Anpassung an komplexe und variable Umweltbedingungen ermöglichen. Damit hat eine Minimalzelle grundsätzlich

---

20 Himmelreich R, Hilbert H, Plagens H, Pirkl E Li BC, Herrmann R; Complete Sequence analysis of the genome of the bacterium Mycoplasma pneumoniae. Nucl. Acids Res., 1996, 24, 4420–4449.

21 Waters E, Hohn MJ, Ahel I, Graham DE, Adams MD, Barnstead M, Beeson KY, Bibbs L, Bolanos R, Keller M, Kretz K, Lin X, Mathur E, Ni J, Podar M, Richardson T, Sutton GG, Simon M, Soll D, Stetter KO, Short JM, Noordewier M; The genome of Nanoarchaeum equitans: insights into early archaeal evolution and derived parasitism. Proc. Natl. Acad. Sci. USA, 2003, 22, 12984–12988.

22 Pérez-Brocal V, Gil R, Ramos S, Lamelas A, Postigo M, Michelena JM, Silva FJ, Moya A, Latorre A; A small microbial genome: the end of a long symbiotic relationship? Science, 2006, 314, 312–313.

23 Nakabachi A, Yamashita A, Toh H, Ishikawa H, Dunbar HE, Moran NA, Hattori M; The 160-kilobase genome of the bacterial endosymbiont Carsonella. Science, 2006, 314, 267.

eine reduzierte Fitness gegenüber Wildtypzellen und eignet sich aus Sicherheitsaspekten besonders für biotechnologische Prozesse und für eine gezielte Freisetzung.

Der *top-down*-Ansatz zur Erzeugung minimaler Genome wurde in der Zwischenzeit bereits bei mehreren Mikroorganismen erprobt, und zwar bei dem Gram-negativen Bakterium *Escherichia coli (E. coli)*[24], bei den Gram-positiven Bakterien *Bacillus subtilis*[25] und *Corynebacterium glutamicum*[26] sowie bei der Hefe *Saccharomyces cerevisiae*[27]. Generell werden zur Genomreduktion nicht essenzielle Gene und intergenische Regionen entfernt. Hierunter fallen zum Beispiel Genregionen, die die Nutzung variabler Nahrungsquellen erlauben oder Elemente für Antworten auf Stresssituationen kodieren. Die Identifizierung von solchen nicht essenziellen Genregionen kann über verschiedene Techniken erfolgen. Als sehr erfolgreich hat sich die Mutationsanalyse erwiesen, wobei u.a. mit Transposons zur Markierung der Mutationsorte gearbeitet wird. Zur gezielten Ausschaltung von Genbereichen durch Deletion ist die annotierte Genomsequenz von ausschlaggebender Bedeutung. Ein interessanter Nebeneffekt ergibt sich dabei aus der systematischen Entfernung von Insertionselementen und Transposons, da hierdurch die für die technische Anwendung wichtige Genomstabilität gesteigert werden kann. Der *top-down*-Ansatz zur Reduktion eines bakteriellen Genoms ist u.a. bei *E. coli* intensiv verfolgt worden. Das *E.-coli*-K-12-Genom konnte unter Beibehaltung der Lebensfähigkeit der Zelle in der Zwischenzeit von 4,6 Mb auf 3,7 Mb reduziert werden.[28]

Der *bottom-up*-Ansatz zur Erzeugung von minimalen Genomen geht vom Entwurf einer Gesamtsequenz eines Minimalgenoms am Reißbrett aus, das nach chemischer Komplettsynthese in eine Zellhülle eingebracht wird und zelluläres Leben ermöglichen soll. Ein solcher *bottom-up*-Ansatz kann ohne Zweifel als ein Herzstück der Synthetischen Biologie betrachtet werden. Allein die Entwicklung von Gesamtsequenzen minimaler Genome am Reißbrett erfordert enormes Wissen über das Zusammenspiel einzelner Zellmodule. Ein solches Zusammenspiel muss mit vielfältigen Methoden der Systembiologie erarbeitet werden. Weitere wichtige Einzelschritte des *bottom-up*-Ansatzes sind jedoch bereits erprobt worden. So gelang der Gruppe um Craig Venter die chemische

24 Pósfai G, Plunkett G, Feher T, Frisch D, Keil GM, Umenhoffer K, Kolisnychenko V, Stahl B, Sharma SS, de Arruda M, Burland V, Harcum SW, Blattner FR; Emergent properties of reduced-genome Escherichia coli. Science, 2006, 312, 1044–1046.

25 Morimoto T, Kadoya R, Endo K, Tohata M, Sawada K, Liu S, Ozawa T, Kodama T, Kakeshita H, Kageyama Y, Manabe K, Kanaya S, Ara K, Ozaki K, Ogasawara N; Enhanced recombinant protein productivity by genome reduction in Bacillus subtilis. DNA Res., 2008, 15, 73–81.

26 Suzuki N, Nonaka H, Tsuge Y, Inui M, Yukawa H; New multiple-deletion method for the Corynebacterium glutamicum genome, using a mutant lox sequence. Appl. Env. Micr., 2005, 71, 8472–8480.

27 Murakami K, Tao E, Ito Y, Sugiyama M, Kaneko Y, Harashima S, Sumiya T, Nakamura A, Nishizawa M; Large scale deletions in the Saccharomyces cerevisiae genome create strains with altered regulation of carbon metabolism. Appl. Micr. Biotechnol. 2007, 75, 589–597.

28 Pósfai G, Plunkett G, Feher T, Frisch D, Keil GM, Umenhoffer K, Kolisnychenko V, Stahl B, Sharma SS, de Arruda M, Burland V, Harcum SW, Blattner FR; Emergent properties of reduced-genome Escherichia coli. Science, 2006, 312, 1044–1046.

Komplettsynthese des 0,583 Mb großen Genoms von *Mycoplasma genitalium*.[29] Dieses Ergebnis kann als wissenschaftlicher Durchbruch betrachtet werden in Anbetracht der Tatsache, dass dieses Genom aus 5 bis 7 kb großen DNA-Stücken *in vitro* und *in vivo* zusammengesetzt wurde. Außerdem wurde bereits gezeigt, dass ein komplettes mikrobielles Genom in eine Zellhülle transplantiert werden kann. Dies gelang mit dem *Mycoplasma-mycoides*-Genom, das sich nach Transplantation in eine *Mycoplasma-capricolum*-Zellhülle als funktionsfähig erwies.[30] Damit sind erste Grundzüge des *bottom-up*-Ansatzes zur Erzeugung von synthetischen Minimalzellen mit Minimalgenomen bereits verwirklicht.

Es stellt sich nun die interessante Frage, welche Größe ein Minimalgenom jeweils haben muss, um bestimmte Lebensvorgänge verschiedener Organismen zu vermitteln. Diese Frage kann nur zufriedenstellend beantwortet werden, wenn für ausgewählte Mikroorganismen sowohl der *top-down*- als auch der *bottom-up*-Ansatz in einer vereinten Strategie verfolgt wird.

3.3 Generierung von Protozellen – Artifizielle Systeme mit Eigenschaften lebender Zellen

Im Gegensatz zu Minimalzellen sind Protozellen keine lebenden Zellen, sondern artifizielle Einheiten. Sie sind im Labor konstruierte, selbst replizierende Nanosysteme, die viele Eigenschaften von lebenden Zellen aufweisen wie zum Beispiel das Vorhandensein eines mutierbaren Informationsspeichers, eines Stoffwechselsystems und einer umhüllenden Membran, die das System abgrenzt, dennoch für den Austausch von Energie und Materie mit der Umgebung selektiv offen ist. Protozellen gelten als Brücke zwischen belebter und unbelebter Materie.[31] Die Synthese von Protozellen soll helfen, die Prinzipien, die Funktionsweisen und die Entstehung von lebenden Zellen zu verstehen. Damit stellt das Design von Protozellen einen Weg zu lernen dar, nach welchen Grundprinzipien eine lebende Zelle tatsächlich funktioniert und entstehen konnte. Diese Fragestellungen ergeben sich ebenso bei der Generierung von Minimalzellen, weshalb in der Literatur oft Minimalzellen mit zu den Protozellen gezählt werden, obwohl die beiden Begriffe nicht im eigentlichen Sinn synonym sind.[32]

29 Gibson DG, Benders GA, Andrews-Pfannkoch C, Denisova EA, Baden-Tillson H, Zaveri J, Stockwell TB, Brownley A, Thomas DW, Algire MA, Merryman C, Young L, Noskov VN, Glass JI, Venter JC, Hutchison CA 3rd, Smith HO; Complete chemical synthesis, assembly, and cloning of a Mycoplasma genitalium genome. Science, 2008, 319, 1215–1220.

30 Lartigue C, Glass JI, Alperovich N, Pieper R, Parmar PP, Hutchison CA 3rd, Smith HO, Venter JC; Genome transplantation in bacteria: changing one species to another. Science, 2007, 317, 632–638.

31 Rasmussen S, Chen L, Deamer D, Krakauer DC, Packard NH, Stadler PF, Bedau MA; Transitions from nonliving to living matter. Science, 2004, 303, 963–965.

32 Rasmussen S, Bedau MA, Chen L, Deamer D, Krakauer DC, Packard NH and Stadler PF (eds.); Protocells. Bridging Nonliving and Living Matter. MIT Press, Cambridge, 2008.

Biobasierte Protozellen werden aus den elementaren Bausteinen von lebenden Zellen konstruiert (DNA, RNA, Proteine, Lipide). Sie können als mögliche Vorläufer von lebenden Zellen angesehen werden. Ein prominentes Beispiel hierfür sind Lipid-Membranvesikel mit eingeschlossenen RNA-Replikationssystemen, die in der Lage sind, Ribonukleotide aufzunehmen und durch Verschmelzung mit im Medium vorhandenen Fettsäure-Mizellen zu wachsen, bis sie sich spontan in zwei „Tochterzellen" teilen.[33,34,35] Bemerkenswert ist auch, dass zellfreie Expressionssysteme (DNA → RNA → Protein) in Lipidmembranvesikel eingebracht werden konnten, was zur Bildung von Nanosystemen führte, die Merkmale lebender Zellen zeigen.[36,37]

Aber auch chemisch-synthetische künstliche Einheiten mit integrierten komplexen elektrischen Schaltkreisen haben sich zu artifiziellen Zellen programmieren lassen, die Funktionen von lebenden Zellen simulieren.[38]

Neben dem Wissensgewinn verspricht die Entwicklung von Protozellen verschiedenster Herstellung interessante angewandte Perspektiven. So könnten synthetisch hergestellte Miniatur-Fabriken für die Produktion von Medikamenten und Feinchemikalien auf der Basis von Protozellen entwickelt werden, eine Option, die allerdings derzeit noch eine Zukunftsvision ist.

Nach dem gegenwärtigen Wissensstand stammen alle heute lebenden Organismen von einem Urzellen-Pool (Progenoten) ab, aus dem sich vor etwa vier Milliarden Jahren auf dieser Erde alles Leben entwickelt hat. Die heutige Wissenschaft ist noch weit davon entfernt, die Evolution des Lebens im Reagenzglas vollständig nachvollziehen zu können und lebende Zellen komplett *de novo* aufzubauen. Aber bereits bei der Synthese von Protozellen wird die Frage angesprochen, wo die Grenzen zwischen toter und lebender Materie liegen und was Leben eigentlich ausmacht. Hierzu gibt es bereits ethische Richtlinien.[39] Ob die Wissenschaft bei dem Versuch, lebende Zellen zu synthetisieren, ethische Grenzen überschreitet, bedarf eingehender Diskussion. Sollte es tatsächlich gelingen, Blaupausen für lebende Zellen mit neuen Eigenschaften zu entwerfen, muss die Wissenschaft diese Frage beantworten. Ausführungen hierzu finden sich im Kapitel 4.4.

---

33 Hanczyc MM, Fujikawa SM, Szostak JW; Experimental models of primitive cellular compartments: Encapsulation, growth and division. Science, 2003, 302, 618–622.

34 Chen IA, Roberts RW, Szostak JW; The emergence of competition between model protocells. Science, 2004, 305, 1474–1476.

35 Mansy SS, Schrum JP, Krishnamurthy M, Tobé S, Treco D, Szostak JW; Template-directed synthesis of a genetic polymer in a model protocell. Nature, 2008, 454, 122–125.

36 Ishikawa K, Sato K, Shima Y, Urabe I, Yomo T; Expression of a cascading genetic network within liposomes. FEBS Lett., 2004, 576, 387–390.

37 Noireaux V, Libchaber A; A vesicle bioreactor as a step toward an artificial cell assembly. Proc. Natl. Acad. Sci. U.S.A., 2004, 101, 17669–17674.

38 McCaskill, JS; Evolutionary microfluidic complementation towards artificial cells. in: Protocells. Bridging Nonliving and Living Matter. eds.: Rasmussen S, Bedau MA, Chen L, Deamer D, Krakauer DC, Packard NH and Stadler PF. MIT Press, Cambridge, 2008, 253–294.

39 Bedau MA, Parke EC, Tangen U, Hantsche-Tangen B; Ethical guidelines concerning artificial cells; www.istpace.org/Web_Final_Report/the_pace_report/Ethics_final/PACE_ethics.pdf

3.4 Design von maßgeschneiderten Stoffwechselwegen

Als ein typisches Beispiel für die Synthetische Biologie wird häufig das Design von maßgeschneiderten Stoffwechselwegen *(metabolic engineering)* angeführt. Im klassischen Sinne versteht man darunter die Modifizierung bzw. Ergänzung vorhandener Biosynthesekapazitäten entweder in bekannten Produktions- oder in Fremdorganismen. Der gewünschte Stoffwechselweg wird in diesem Fall mit Regelschaltkreisen und Integrationsmodulen auf dem Reißbrett entworfen. Die dazu erforderlichen DNA-Sequenzen werden chemisch synthetisiert, zusammengefügt (rekombiniert) und anschließend in einen geeigneten Empfängerorganismus transferiert.

Der gezielte Transfer einzelner Gene in fremde Wirtsorganismen, wie das Bakterium *Escherichia coli*, die Hefe *Saccharomyces cerevisiae* oder selbst in Humanzellen ist seit den 1970er-Jahren gängige Laborpraxis. Dieser Ansatz, der bereits den Transfer von DNA in einem Umfang von mehreren zehntausend Basenpaaren umfasst, ist besonders im Bereich der Antibiotika- und Aminosäureherstellung oder auch in der Entwicklung transgener Pflanzen beschritten worden.[40] Im Vordergrund steht dabei die Optimierung des Synthesepotenzials eines Produktionsstammes. Somit wird beim *metabolic engineering* wissenschaftliches Interesse mit einer kommerziellen Anwendung kombiniert.[41]

In jüngerer Zeit konnten auch Wege für artifizielle und in der Natur in dieser Form nicht vorkommende neuartige Biosyntheseprozesse eröffnet werden. Bei dieser Vorgehensweise handelt es sich allerdings weniger um eine neue Technologie als um eine Weiterentwicklung des *metabolic engineering*, wie es seit Mitte der 1980er-Jahre bekannt ist. Handelte es sich bis dahin um die gezielte Veränderung einzelner Gene oder ihrer Regulatoren in einem mehrere Gene umfassenden Biosynthese-Gencluster, gelang 2003 die gentechnische Konstruktion eines kompletten Biosyntheseweges für Isoprenoide in *E. coli*. Dieses Bakterium wurde so programmiert, dass es eine Vorstufe des Antimalaria-Medikaments Artemisinin, die Artemisinsäure, synthetisiert.[42] Dabei wurden Gene aus der Pflanze *Artemisia anna*, der Bäckerhefe sowie bakterielle Gene in *E. coli* zusammengesetzt und mit den notwendigen bakteriellen Kontrollregionen für eine regulierte Genexpression versehen. Drei Jahre später konnte auch die Hefe zum Artemisinsäureproduzenten programmiert werden.[43] Bei der Umsetzung des Verfahrens arbeiten derzeit die Non-Governmental Organisation (NGO) One World Health, das Biotechnologieunternehmen Amrys, die Bill Gates

---

40 Rodriguez E, McDaniel R; Combinatorial biosynthesis of antimicrobials and other natural products. Curr. Opin. Microbiol., 2001, 4, 526–534.

41 Durot M, Bourguignon PY, Schachter V; Genome-scale models of bacterial metabolism: reconstruction and applications. FEMS Microbiol. Rev., 2009, 33, 164–190.

42 Martin VJ, Pitera DJ, Withers ST, Newman JD, Keasling JD; Engineering a mevalonate pathway in Escherichia coli for production of terpenoids. Nat. Biotechnol. 2003, 21, 796–802.

43 Ro DK, Paradise EM, Ouellet M, Fisher KJ, Newman KL, Ndungu JM, Ho KA, Eachus RA, Ham TS, Kirby J, Chang MC, Withers ST, Shiba Y, Sarpong R, Keasling JD; Production of the antimalarial drug precursor artemisinic acid in engineered yeast. Nature, 2006, 440, 940–943.

Foundation sowie das Pharmaunternehmen Sanofi-Aventis zusammen. Das Ziel dieser Arbeiten ist die Herstellung des Malariamittels, um es für Patienten in Ländern, in denen die Malaria endemisch ist, kostengünstig verfügbar zu machen.

Ein weiteres Beispiel ist die Synthese von Hydrocortison aus Ethanol in der Bäckerhefe. 2003 gelang dieses Verfahren durch das funktionale Zusammenschalten von 13 Genen, von denen acht menschlichen Ursprungs sind. Auch hier steht die preisgünstige Herstellung des Produkts im Vordergrund.[44] Im Vergleich zur herkömmlichen Totalsynthese von Hydrocortison, die bis zum Endprodukt über 23 chemische und biotechnologische Reaktionsschritte verläuft, stellt dieses Verfahren einen wichtigen Fortschritt im Produktionsverfahren dar.[45]

Neben den oben genannten Arbeiten, die im Bereich der pharmazeutischen Entwicklung angesiedelt sind, gewinnt die Konstruktion synthetischer Gencluster bzw. artifizieller Biosynthesewege auch im Umfeld der industriellen „weißen" Biotechnologie zunehmend an Bedeutung. Es wird unter anderem angestrebt, petrochemische Herstellungsverfahren durch nachhaltige Bioverfahren unter Verwendung nachwachsender Rohstoffe zu ersetzen. Beispielhaft hierfür steht die Bildung eines Ausgangsstoffes für die Herstellung von Nylon.[46]

Einen bemerkenswerten Fortschritt stellt der Transfer umfangreicher Gencluster, die für neue Naturstoffe kodieren, in fremde Wirtsbakterien dar. Darüber hinaus lassen sich „stumme Gencluster" in ihrer Expression aktivieren. Zum Beispiel gelang die funktionale Expression eines Genclusters für die Bildung eines Naturstoffes aus dem Myxobakterium *Stigmatella* in *Pseudomonas*. Dadurch wurde auch die Möglichkeit eröffnet, den Naturstoff gezielt zu verändern und erheblich verbesserte Produktausbeuten zu erzielen.[47,48] Ferner gibt es inzwischen verbesserte DNA-Transfersysteme, die eine Klonierung von Genclustern > 80 kb in *E. coli* und deren Expression in anderen Wirtsorganismen, zum Beispiel *Streptomyces lividans,* erlauben. Dieses konnte jüngst für das Polyketidantibiotikum Meridamycin demonstriert werden.[49]

44 Szczebara FM, Chandelier C, Villeret C, Masurel A, Bourot S, Duport C, Blanchard S, Groisillier A, Testet E, Costaglioli P, Cauet G, Degryse E, Balbuena D, Winter J, Achstetter T, Spagnoli R, Pompon D, Dumas B; Total biosynthesis of hydrocortisone from a simple carbon source in yeast. Nat. Biotechnol., 2003, 21, 143–149.

45 Redaktion PROCESS; Hefezelle als Wirkstofffabrik. PROCESS, 22.02.2007; www.process.vogel.de/articles/58824/

46 Niu W, Draths KM, Frost JW; Benzene-free synthesis of adipic acid. Biotechnol. Prog., 2002, 18, 201–211.

47 Wenzel SC, Gross F, Zhang Y, Fu J, Stewart AF, Müller R; Heterologous expression of a myxobacterial natural products assembly line in pseudomonads via red/ET recombineering. Chem. Biol., 2005, 12, 349–356.

48 Perlova O, Gerth K, Kuhlmann S, Zhang Y, Müller R; Novel expression hosts for complex secondary metabolite megasynthetases: Production of myxochromide in the thermopilic isolate Corallococcus macrosporus GT-2. Microb. Cell Fact., 2009, 8, 1–11.

49 Liu H, Jiang H, Haltli B, Kulowski K, Muszynska E, Feng X, Summers M, Young M, Graziani E, Koehn F, Carter GT, He M; Rapid cloning and heterologous expression of the meridamycin biosynthetic gene cluster using a versatile Escherichia coli-Streptomyces artificial chromosome vector, pSBAC (perpendicular). J. Nat. Prod., 2009, 72, 389–395.

Die Reihe der ausgewählten Beispiele ließe sich weiter ergänzen. Ihnen ist gemein, dass sie auf der detaillierten Kenntnis der Biosynthesewege, einem rationalen Konzept („rational design") und der Weiterentwicklung des gentechnisch experimentellen Methodenrepertoires beruhen. Basierend auf der DNA-Sequenzierung einer Vielzahl von Biosynthesegenclustern und der Aufklärung der zugrunde liegenden Expressionskontrollen wird es in Zukunft üblich sein, die DNA statt auf dem zeitaufwendigen, klassischen Klonierungsweg auch kostengünstig synthetisch herzustellen (s. Kap. 3.1). Darüber hinaus ist die Möglichkeit gegeben, die genetische Information dem jeweiligen Produktionswirt optimal anzupassen. Hier bietet sich ein bislang noch nicht hinreichend ausgeschöpftes Anwendungspotenzial an.

Inwieweit beim *metabolic engineering* auch synthetisch konstruierte Produktionswirte wie Minimalzellen und Protozellen zum Einsatz kommen werden, ist von deren Produktbildungskapazitäten abhängig.

3.5 Konstruktion von komplexen genetischen Schaltkreisen

Seit der Beschreibung genetischer Schaltkreise durch Jacob und Monod[50] in den 1960er-Jahren sind Molekularbiologen daran interessiert, die vielfältigen Möglichkeiten zu nutzen, um zelluläre Regulationsvorgänge zu modifizieren und in extern kontrollierbare, genetische Schaltkreise zu überführen.

DNA entwickelt ihre biologische Funktion erst über die exakte Steuerung der Genaktivität. Viren, Bakterien und eukaryontische Zellen nutzen hierzu eine Fülle komplexer Regelmechanismen, die auf der Ebene der Nukleinsäuren als regulatorische Motive niedergelegt sind und in Wechselwirkung mit zellulären Faktoren (RNAs oder Proteine) treten. Die Genaktivität kann so auf allen Ebenen der Genexpression – von der Bildung des Primärtranskripts über die (in Eukaryonten anzutreffende) post-transkriptionelle Modifikation bis hin zur Proteinbiosynthese – fein auf die metabolischen und gewebespezifischen Anforderungen des Zellhaushalts abgestimmt werden.

Das heute in der Biotechnologie am häufigsten eingesetzte künstliche Regelsystem nutzt sogenannte Tetracyclin-sensitive Promotoren. Diese beruhen auf der Adaptation eines bakteriellen Antibiotika-Spürsystems für die kontrollierte Genexpression in Zellen. Tetracyclin-sensitive Promotoren spielen bereits seit vielen Jahren eine große Rolle in der Funktionsanalyse von Genen und sind auch für die biotechnologische Produktherstellung oder die therapeutischen Anwendungen im Sinne einer somatischen Gentherapie interessant.[51]

---

50 Jacob F, Monod J; Genetic regulatory mechanisms in the synthesis of proteins. J. Mol. Biol., 1961, 3, 318–356.

51 Goosen M, Bujard H; Studying gene function in eukaryotes by conditional gene inactivation. Annu. Rev. Genet., 2002, 36, 153–173.

Eine Vielzahl weiterer genetischer Schaltkreise wurde in den vergangenen Jahren in Zellen eingeführt, die neben der transkriptionellen Kontrolle auch post-transkriptionelle Mechanismen ansteuern; auch das Tetracyclin-regulierte System bleibt weiterhin Gegenstand umfangreicher Optimierungen.[52] Werden nun mehrere solcher Schaltkreise kombiniert, können über positive und negative Rückkopplungen komplexe kybernetische Systeme unterschiedlicher Ausprägung entstehen. Eine paradigmatische Rolle spielt der sogenannte Repressilator, ein oszillierendes regulatorisches System, das auf der Kombination von drei bakteriellen Repressorproteinen beruht.[53] Die Konstruktion noch komplexerer genetischer Schaltkreise wird in zunehmendem Maße von der Entwicklung funktionell definierter Module im Sinne der „BioBricks" profitieren. Ihr Zusammenspiel ist nur bedingt berechenbar und muss daher empirisch überprüft werden.[54,55] Insofern sind die Grenzen zwischen klassischer Biotechnologie und Synthetischer Biologie bei der Entwicklung künstlicher Schaltkreise fließend.

Vom Grundsatz her sollte die Abhängigkeit von Organismen mit künstlichen genetischen Schaltkreisen in ihrer Regulation von exogen applizierbaren Pharmaka bzw. anderen Formen chemisch oder physikalisch definierter Induktoren die biologische Sicherheit erhöhen.

3.6 Schaffung von orthogonalen Biosystemen

Bei der Konstruktion neuartiger Biosysteme spielt die Komplexität eine zentrale Rolle: Neu eingebrachte Moleküle oder Schaltkreise interagieren mit dem bestehenden System. Um möglichst unabhängig voneinander funktionierende Bausteine zu integrieren, verfolgt man das Konzept orthogonaler Biosysteme. Ein möglicher Ertrag ist eine Verbesserung der biologischen Sicherheit.

Orthogonalität bedeutet in diesem Zusammenhang die freie Kombinierbarkeit unabhängiger Bauteile und ist ein technikwissenschaftliches Konstruktionsprinzip, das unter anderem in der Informatik eine wichtige Rolle spielt. Die mit Orthogonalität verbundene Strategie hat zum Ziel, Teilsysteme zu verändern, ohne gleichzeitig andere Teilsysteme erheblich zu stören. Die Verwirklichung von Orthogonalität in biologischen Systemen wird als Voraussetzung für eine Synthetische Biologie im Sinne gezielter Eingriffe gesehen, die über den rein empirischen Ansatz hinausgehen und die nicht in der zellulären Komplexität gefangen sind.[56]

52 Greber D, Fussenegger M; Mammalian synthetic biology: engineering of sophisticated gene networks. J. Biotechnol., 2007, 130, 329–345.

53 Elowitz MB, Leibler S; A synthetic oscillatory network of transcriptional regulators. Nature, 2000, 403, 335–338.

54 Stricker J, Cookson S, Bennett MR, Mather WH, Tsimring LS, Hasty J; A fast, robust and tunable synthetic gene oscillator. Nature, 2008, 456, 516–519.

55 Tigges M, Marquesz-Lago TT, Stelling J, Fussenegger M; A tunable synthetic mammalian oscillator. Nature, 2009, 457, 309–312.

56 Panke S; Synthetic Biology – Engineering in Biotechnology. 2008, Swiss Academy of Technical Sciences (Ed.).

Um unabhängig voneinander funktionieren zu können, sollten orthogonale Teilsysteme möglichst „unsichtbar" für den Rest der Zelle sein, also deren Wechselwirkung mit den natürlichen (Teil-)Systemen minimal beeinflussen.

Ein Beispiel ist das Engineering des genetischen Codes: Die Proteine sind in der Regel aus 20 verschiedenen Aminosäuren aufgebaut, die deren Struktur und Funktion prägen. Es gibt freilich keinen chemischen oder biologischen Grund, warum nicht andere als die 20 „kanonischen" Aminosäuren als Bausteine für Proteine biologische Verwendung finden könnten. Um künstliche Aminosäuren an ausgewählten Positionen eines Proteins einzuschleusen, können beispielsweise Kodons modifiziert und die zelluläre Translationsmaschinerie entsprechend angepasst werden, die genetische Information wird dann am Ribosom anders übersetzt.

Ein Ansatz, den genetischen Code gezielt für eine künstliche Aminosäure zu erweitern, basiert darauf, das am wenigsten verwendete Stopp-Kodon für den Einbau dieser Aminosäure zu verwenden. Hierzu müssen eine entsprechend modifizierte Transfer-RNA (tRNA) und das Beladungsenzym in die Zelle eingebracht werden. Idealerweise erkennt diese tRNA ausschließlich das Stopp-Kodon und fügt bei der ribosomalen Proteinsynthese hierfür die zusätzliche Aminosäure ein, ohne dass die Wirkung der bereits vorhandenen tRNAs berührt wird.[57]

Ein anderes Beispiel für ein orthogonales System ist ein verändertes Ribosom, das ein Leseraster aus Quadrupletts, das heißt aus vier statt den üblichen drei Basen je Kodon, bearbeitet.[58] Ziel ist es, zwei unabhängig voneinander arbeitende Übersetzungssysteme in einer Zelle zu etablieren: ein „natürliches" zur Synthese normaler Zellproteine und ein „orthogonales" zur Synthese von Polymeren aus nicht natürlich vorkommenden Aminosäuren. Auf diese Weise könnten lebende Zellen zur Synthese beliebiger Aminosäurepolymere programmiert werden, die als neue Werkstoffe (Zahnimplantate, Knorpel- und Knochenersatz), als therapeutische Wirkstoffe und für Forschungszwecke zur Struktur- und Funktionsaufklärung dienen könnten.

Orthogonale Biosysteme stellen eine Erhöhung der biologischen Sicherheit in Aussicht. So können zum Beispiel Gene, die über einen nicht natürlichen genetischen Code für die Synthese eines bestimmten Genprodukts programmiert sind, ausschließlich in Organismen mit diesem orthogonalen Translationssystem entschlüsselt werden (s. Kap. 4.3).

---

57 Budisa N, Weitze MD; Den Kode des Lebens erweitern. Spektrum der Wissenschaft, Januar 2009, 42–50.

58 Wang K, Neumann H, Peak-Chew SY, Chin JW; Evolved orthogonal ribosomes enhance the efficiency of synthetic genetic code expansion. Nat. Biotechnol., 2007, 25, 770–777.

4   Aktuelle Herausforderungen

4.1 Ökonomische Aspekte

4.1.1 Marktpotenziale

Die ökonomischen Aussichten der Synthetischen Biologie lassen sich an den kommerziellen Verwendungsmöglichkeiten im industriellen und medizinischen Bereich messen sowie an Lizenzeinnahmen und am Schutz des geistigen Eigentums durch Patente ablesen. Wenn sich die Synthetische Biologie bislang auch noch weitgehend im Forschungsstadium befindet, so zeichnen sich bereits jetzt attraktive Marktpotenziale ab. Dabei liegen die ökonomisch interessanten Möglichkeiten in der erhöhten Produktivität durch die Verbesserung von Herstellungsprozessen, der Gewinnung neuer Produkte, der Beschleunigung von Entwicklungszeiten durch Standardisierung biologischer Bauteile und Etablierung neuer Produktionskonzepte. Hohe Marktpotenziale bezogen auf den Produktionsstandort Deutschland sind vor allem im Bereich der Weißen Biotechnologie, der Bioenergie sowie in der Medizin zu erwarten. Neue Produktionsverfahren zeichnen sich durch die Schaffung bislang nicht bekannter Synthesewege ab, und Möglichkeiten werden eröffnet, Produktionsstämme mit verbesserten Eigenschaften zu konstruieren. Zudem entwickeln sich Dienstleistungen im Bereich der Analyse und Herstellung von Nukleinsäuren, die auf Technologien zurückgreifen, die bereits unter Patentschutz stehen.

Die in Deutschland traditionell starke chemische Industrie nutzt bereits heute vielfältige Verfahren der Weißen Biotechnologie. Hieran lässt sich erkennen, welches Potenzial in der Schaffung neuer Prozesse mittels Synthetischer Biologie liegt. Diese Prozesse könnten neue Rohstoffquellen nutzen, natürliche Ressourcen sparen helfen und Abfälle vermeiden. Zum Beispiel wird die als Futtermittelzusatz benötigte Aminosäure Lysin derzeit mit klassischen biotechnologischen Verfahren im Maßstab von 700 000 Tonnen jährlich produziert, was einem Marktwert von 1,4 Milliarden Euro entspricht. In Anbetracht dieses hohen Umsatzes können schon kleinste Optimierungen in dem biotechnologischen Verfahren erhebliche wirtschaftliche Relevanz haben. Deshalb hat das *metabolic engineering* in dem Marktkonzept eine beachtliche Bedeutung (s. Kap. 3.4).

Mit der Umstellung einer auf fossilen Rohstoffen basierten Produkt- und Energiewirtschaft auf erneuerbare Ressourcen gibt es zukünftig zwei Ansätze zur konzeptionellen Umstellung dieser Industriezweige. Aus ökonomischen Erwägungen ist es zunächst sinnvoll, heute verwendete Ausgangsverbindungen auf der Basis nachwachsender Rohstoffe zu produzieren, da auf diese Weise bestehende Produktionsanlagen weiter genutzt werden können. Mittelfristig ist ein Ersatz petrochemischer Ausgangsverbindungen durch biologisch leicht zugängliche Substanzen anzustreben, was eine schrittweise Umstellung der Produktionsverfahren und -anlagen zur Folge hätte.[59]

Die Synthetische Biologie verspricht auch neue Strategien zur Gewinnung von Biokraftstoffen. Biokraftstoffe der ersten Generation basieren auf Pflanzen, die auch als Nahrungsmittel dienen. Angesichts begrenzter Kapazitäten der Agrarflächen entsteht so eine Spannung zum Nahrungsmittelanbau. Verfahren zur Herstellung von Biokraftstoffen der zweiten Generation nutzen die ganze Pflanze, also insbesondere Teile, die als Nahrungsmittel nicht infrage kommen. Solche Verfahren, bei denen beispielsweise Ethanol aus Agrarabfällen und pflanzlichen Reststoffen gewonnen wird, könnten durch die Synthetische Biologie beflügelt werden. Auch die Gewinnung von Bio-Wasserstoff aus Wasser und Sonnenenergie könnte langfristig mithilfe maßgeschneiderter Mikroorganismen oder biomimetisch konzipierter Katalysatoren ein technisch durchaus realisierbares Verfahren werden. Forschungen auf diesen Gebieten werden durch große Ölkonzerne und durch die Energiewirtschaft aufmerksam verfolgt und teilweise unterstützt.

Vielfältige Marktpotenziale bieten sich für die Synthetische Biologie im Bereich der medizinischen Diagnostik und Prävention, der Arzneimittelentwicklung sowie dem Einsatz alternativer Therapieverfahren an. Auf mögliche Anwendungen im Bereich der Medizin, der Arzneimittelentwicklung und der Wirkstoffproduktion wurde in den Kapiteln 3.1 und 3.4 bereits hingewiesen.

Um neue Marktpotenziale wirtschaftlich gewinnbringend zu erschließen und um den Transfer des Grundlagenwissens in die Anwendung zu beschleunigen, sind eine weitere Stärkung der interdisziplinären Arbeitsweise und eine frühe Beteiligung der ingenieurwissenschaftlichen Fachrichtungen erforderlich.

4.1.2 Patentrechtliche Fragen

Gene und Genfragmente, die für eine bestimmte Funktion kodieren, lassen sich patentieren, in Europa geregelt durch die EU-Richtlinie 98/44/EC[60] und deren Implementierung in das Europäische Patentübereinkommen. Dies trifft auch auf synthetische Elemente zu, die teilweise als „BioBricks" bezeichnet werden. Für Aufsehen haben 2007 US-amerikanische und internationale Patentanmeldungen des J. Craig Venter Institute gesorgt, in dem exklusive Eigentumsrech-

---

59   Eine starke Biologisierung der Wirtschaft im Rahmen einer Bioökonomie wird prognostiziert: Man erwartet, dass Biomaterialien und Bioenergie bis 2030 ein Drittel der Industrieproduktion in Europa ausmachen werden, vgl. „En Route to the Knowledge-based Bio-Economy", Cologne Paper, Mai 2007.

60   Directive 98/44/EC of the European Parliament and of the Council of 6 July 1998 on the legal protection of biotechnological inventions; http://eur-lex.europa.eu/LexUriServ/LexUriServ.do?uri=OJ:L:1998:213:0013:0021:EN:PDF

te an mehreren essenziellen Genen von *Mycoplasma* und einem synthetischen Organismus (*Mycoplasma laboratorium*) angemeldet wurden, der mithilfe dieser Gene wachsen und sich eigenständig replizieren können soll. Die Tür zur Sicherung von Eigentumsrechten an gentechnisch veränderten Organismen (GVO) wurde bereits 1980 durch eine Entscheidung des amerikanischen Obersten Gerichtshofs aufgestoßen, der im Fall Chakrabarty befand, dass ein GVO nicht als Produkt der Natur angesehen werden kann und daher grundsätzlich, das heißt, sofern weitere Voraussetzungen (zum Beispiel Neuheitswert) erfüllt sind, patentierbar ist.[61] In Europa sind mikrobiologische Verfahren und die mithilfe dieser Verfahren gewonnenen Erzeugnisse grundsätzlich patentierbar (Art. 53b) EPÜ). Ebenso ist biologisches Material, das mithilfe eines technischen Verfahrens aus seiner natürlichen Umgebung isoliert oder hergestellt wird, auch wenn es in der Natur schon vorhanden war, patentierbar (Regel 27a) EPÜ).

Die Patentierung von GVO gewährt demjenigen, der – etwa aufgrund aufwendiger Forschung und durch geistige Leistung – eine Erfindung gemacht hat, einen Marktvorsprung, indem er andere auf Zeit von der gewerblichen Benutzung der patentierten Erfindung ausschließen oder sie ihnen gegen Lizenzen gestatten kann. Zudem fördern Patente die wissenschaftliche Entwicklung dadurch, dass die Erfindung so deutlich und vollständig zu offenbaren ist, dass ein Fachmann sie ausführen kann; damit werden der Öffentlichkeit Kenntnisse zur Verfügung gestellt, auf deren Grundlage Weiterentwicklungen und Verbesserungen stattfinden können. Jedoch wird auf die Gefahr einer Monopolstellung auf synthetische Organismen verwiesen, die zu einer Vormachtstellung einzelner Unternehmen führen könnte.[62] Dies kann insbesondere kritisch sein, wenn sich bestimmte Plattformtechnologien als Standard oder *de-facto*-Standard etablieren. Befürchtet wird ein mangelnder Zugang zu gesellschaftlich wichtigen Forschungsmaterialien und Anwendungsmöglichkeiten, falls entsprechende Patente zu weit gefasst sind. Ein weiteres Problem könnte die Entstehung von sogenannten „Patent thickets (Patentdickichten)", wie sie aus der Elektronikindustrie bekannt sind, darstellen.[63] Da für die Synthetische Biologie oft eine große Anzahl von „Bausteinen" benötigt wird, könnte die Existenz zahlreicher Rechte an diesen Bausteinen, die möglicherweise von verschiedenen Rechteinhabern gehalten werden, die Entwicklung neuer Produkte erschweren.[64] Um einen solchen Trend zu verhindern, wird von einigen Organisationen, wie der gemeinnützigen BioBricks Foundation, Wert auf frei zugängliche Ressourcen für die Synthetische Biologie gelegt. Die Stiftung hat sich insbesondere zum Ziel gesetzt, DNA-Bausteine, mit denen Biosynthesesysteme zusammengesetzt wer-

---

61   Diamond vs. Chakrabarty, 447 U.S. 303 (1980), US Supreme Court;
     http://caselaw.lp.findlaw.com/scripts/getcase.pl?navby=CASE&court=US&vol=447&page=303

62   Siehe zum Beispiel ETC Group (Action group on Erosion, Technology and Concentration);
     Extreme Genetic Engineering: An Introduction to Synthetic Biology. 2007, 1–64.

63   Shapiro C; Navigating the Patent Thicket: Cross Licenses, Patent Pools, and Standard Setting.
     Innovation Policy and the Economy, 2000, 1, 119–150.

64   Henkel J, Maurer SM; The economics of synthetic biology. Mol. Syst. Biol., 2007, 3, 117.

den können, der Öffentlichkeit frei zugänglich zu machen.[65] Es ist allerdings nicht immer ersichtlich, ob nicht doch gewisse Einzelbestandteile der zur Verfügung gestellten „BioBricks" bereits anderweitig patentrechtlich geschützt sind.

Von weiten Patenten kann eine mittelbare Behinderung der Forschung insofern ausgehen, als kommerzielle Unternehmen wenig geneigt sind, in Forschungsbereiche zu investieren, deren spätere anwendungsbezogene Umsetzung bereits umfassend von Patenten erfasst ist. Auch eine unmittelbare Behinderung der Forschung ist nicht von der Hand zu weisen. Handlungen zu Versuchszwecken, die sich auf den Gegenstand der patentierten Erfindung beziehen, sind nach § 11 Nr. 2 PatG ausdrücklich von der Wirkung des Patents ausgenommen. Gleiches gilt für die Nutzung biologischen Materials zum Zweck der Züchtung, Entdeckung und Entwicklung einer neuen Pflanzensorte (§ 11 Nr. 2a PatG) sowie für Studien und Versuche sowie die sich daraus ergebenden praktischen Anforderungen, die für die Erlangung einer arzneimittelrechtlichen Genehmigung für das Inverkehrbringen in der Europäischen Union oder einer arzneimittelrechtlichen Zulassung in den Mitgliedstaaten der Europäischen Union oder in Drittstaaten erforderlich sind (§ 11 Nr. 2b PatG). Das Versuchsprivileg findet seine Grenze unter anderem darin, dass Versuche nur dann unschädlich sind, wenn sie den patentierten Gegenstand als Objekt der Untersuchung nutzen und nicht lediglich als ein Mittel zu deren Durchführung.

4.2 Forschungsförderung und Ausbildung

Die Synthetische Biologie ist seit etwa 2003 in den Blickpunkt der Forschungsförderung geraten. Inzwischen gibt es eine Reihe nationaler Schwerpunkte, zum Beispiel in Großbritannien, Dänemark, in den Niederlanden und der Schweiz sowie in Frankreich und Deutschland, wofür beispielhaft der im Rahmen der Exzellenzinitiative von der DFG geförderte Exzellenzcluster „bioss" (Biological Signalling Studies) der Universität Freiburg steht, der die Methoden der Synthetischen Biologie mit Studien zur biologischen Signalübertragung verbindet.

Von den einzelnen europäischen Fördermaßnahmen, die gezielt Themen der Synthetischen Biologie zum Inhalt haben, werden hier nur einige beispielhaft aufgeführt. Bereits im 6. Rahmenprogramm der Europäischen Kommission wurden von 2007 bis 2008 innerhalb der „NEST (New and Emerging Science and Technology) Pathfinder Initiative" 18 Projekte mit einem Volumen von 24,7 Millionen Euro gefördert. Darunter befanden sich nicht nur Vorhaben, die auf die Entwicklung neuer Produkte und Methoden ausgerichtet waren, sondern auch Projekte zur Forschungskommunikation (SynBioComm), Fragen der biologischen Sicherheit und ethische Aspekte (SYNBIOSAFE) sowie strategische Planungen (TESSY – Towards a European Strategy for Synthetic Biology). Es ist davon auszugehen, dass der NEST-Initiative, die 2008/09 ausläuft, neue Projekte im 7. Rahmenprogramm folgen werden. Darüber hinaus wurde von

---

65 http://bbf.openwetware.org/

der Europäischen Kommission von 2004 bis 2008 das integrierte Projekt „Pro-gammable Artificial Cell Evolution" (PACE) gefördert. Die Projektgruppe hat „Ethical guide lines concerning artificial cells" herausgegeben, die den derzeiti-gen Stand der Diskussion für dieses Teilgebiet wiedergeben.[66]

Auch die European Science Foundation (ESF) hat besondere Förderprogram-me im Rahmen der Synthetischen Biologie aufgelegt, zum Beispiel eine Aus-schreibung zum EuroCore EuroSYNBIO (Synthetic Biology: Engineering Com-plex Biological Systems). Die Mittel für dieses Programm kommen von den jeweiligen beteiligten nationalen Förderorganisationen, in Deutschland von der DFG. Neben diesen koordinierten Aktivitäten werden auch die themenoffenen Förderverfahren der DFG, zum Beispiel die Einzelförderung, für Projekte aus dem Bereich der Synthetischen Biologie genutzt.

Dieser Überblick zeigt, dass zahlreiche Förderinstrumente zur Forschung auf dem Gebiet der Synthetischen Biologie verfügbar und bei Bedarf ausbaufähig sind. Der Erfolg all dieser Fördermaßnahmen wird jedoch maßgeblich davon abhängen, inwieweit es gelingen wird,

▶ die unterschiedlichen fachlichen Disziplinen zusammenzuführen, um Syn-ergien zu erzeugen;

▶ die vorhandenen Infrastrukturen optimal zu nutzen und durch konzertierte Maßnahmen effizient zu ergänzen;

▶ weitsichtig die Grundlagenforschung zu berücksichtigen, da sich noch viele Gebiete der Synthetischen Biologie auf der Ebene des elementaren Erkennt-nisgewinns bewegen;

▶ zugleich frühzeitig den Anwendungsaspekt in die strategische Planung ein-zubeziehen, um eine schnellere Transformation in die industrielle Nutzung zu erwirken;

▶ durch Information und Kommunikation eine Transparenz zu schaffen, die zur Akzeptanz dieser Forschungsrichtung in der Öffentlichkeit beiträgt.

Schließlich wird der Erfolg der Synthetischen Biologie von der Qualifikation, dem Ideenreichtum und der Motivation junger Nachwuchswissenschaftlerin-nen und Nachwuchswissenschaftler abhängig sein.

Um der letztgenannten Voraussetzung gerecht zu werden, ist es erforderlich, Aspekte der Synthetischen Biologie in den Ausbildungsplänen von Naturwis-senschaftlerinnen und Naturwissenschaftlern und Ingenieurinnen und Inge-nieuren zu verankern. Die Bachelor- und Masterstudiengänge in Europa und eine zunehmende Zahl von Graduiertenkollegs und Doktorandenakademien bieten hierzu Möglichkeiten, die bisher nicht in einem wünschenswerten Um-fang genutzt werden. So sollten Biologinnen und Biologen bereits zu einem frühen Zeitpunkt die Möglichkeit erhalten, ihre grundlegenden Kenntnisse in Chemie, Physik und Mathematik zu vertiefen, um ihre Fähigkeit zum quantita-tiven Denken zu stärken. Andererseits sollten auch Naturwissenschaftlerinnen

---

66   Bedau MA, Parke EC, Tangen U, Hantsche-Tangen B; Ethical guidelines concerning artificial cells; www.istpace.org/Web_Final_Report/the_pace_report/Ethics_final/PACE_ethics.pdf

und Naturwissenschaftler aus nicht lebenswissenschaftlichen Disziplinen sowie Ingenieurinnen und Ingenieure Einblicke in die Physiologie und Biochemie lebender Organismen und in die Techniken der Molekularbiologie erhalten. Dies ist für eine gemeinsame Sprachfindung, ein konzertiertes Vorgehen und ein produktives Handeln unerlässlich.

Diese Vorgehensweise könnte in einem frühen Stadium der Ausbildung gezielt Interessen wecken, junge Wissenschaftlerinnen und Wissenschaftler für eine interdisziplinäre Arbeit begeistern und die Bereitschaft zur Teamarbeit fördern. Eine Möglichkeit zur Motivation bietet unter anderem der seit 2003 stattfindende Wettbewerb iGEM (international Genetically Engineered Machine Competition), bei dem Arbeitsgruppen aus der ganzen Welt ihre Ideen im Rahmen der Synthetischen Biologie einer kritischen Jury präsentieren.

Schließlich sollten den Studienabsolventinnen und -absolventen, die einen anspruchsvollen Ausbildungsweg durchlaufen haben, auch attraktive berufliche Aussichten sowohl im akademischen als auch industriellen Bereich geboten werden.

4.3 Sicherheitsfragen

Die meisten der im Kapitel 3 aufgeführten Forschungsrichtungen der Synthetischen Biologie verwenden molekularbiologische Methoden der Gentechnik. Über die Gentechnik hinaus wird durch die Umsetzung ingenieurwissenschaftlicher Prinzipien in der Synthetischen Biologie ein neuer Aspekt eingeführt.[67,68,69,70] Dieser Ansatz führt nach Meinung einiger Wissenschaftlerinnen und Wissenschaftler weg vom bisherigen Analysieren und Modifizieren, hin zum Synthetisieren und Konstruieren in der Synthetischen Biologie.[71] Das Ziel der Synthetischen Biologie, Genome *in vitro* zu synthetisieren und neuartige Organismen ohne Referenz in der Umwelt zu kreieren, stellt an die biologische Sicherheit in Laboratorien oder bei Freisetzungen (Biosafety) bisher keine zusätzlichen Anforderungen und birgt hinsichtlich der Missbrauchsmöglichkeiten (Biosecurity) dieser Technologie aus heutiger Sicht keine andersartigen Risiken als die Gentechnik. Eine gesetzliche Regulierung speziell für die Synthetische Biologie ist derzeit aus diesen Gründen nicht erforderlich.

Aufgrund der schnellen Entwicklung wird zum jetzigen Zeitpunkt jedoch ein Monitoring der Arbeiten auf dem Gebiet der Synthetischen Biologie durch die ZKBS (Zentrale Kommission für die Biologische Sicherheit) empfohlen und die

---

67   Forum Genforschung; Synthetic Biology. 2007, Platform of the Swiss Academy of Science.

68   Benner SA, Sismour AM; Synthetic Biology. Nat. Rev. Genet., 2005, 6, 533–543.

69   Heinemann M, Panke S; Synthetic Biology – putting engineering into biology. Bioinformatics, 2006, 22, 2790–2799.

70   Keasling JD; Synthetic biology for synthetic chemistry. ACS Chem. Biol., 2008, 3, 64–76

71   van Est R, de Vriend H, Walhout B; Constructing Life. The World of Synthetic Biology. The Hague, Rathenau Institute. 2007, 1–16.

Einrichtung einer behördlichen Kontaktstelle für Unternehmen aus dem Bereich der *in-vitro*-Synthese von Nukleinsäuren vorgeschlagen. Diese Kontaktstelle sollte den Unternehmen Informationen zum Risikopotenzial einzelner Nukleinsäuren zur Verfügung stellen können. Die Einrichtung einer wissenschaftlich fundierten und international abgestimmten Datenbanklösung erscheint hierfür notwendig (s. Kap. 4.3.3).

4.3.1 Biologische Sicherheit (Biosafety)

Biologische Systeme unterliegen dem Einfluss vielfältiger Signale, die über Signalkomponenten – ähnlich einem elektronischen Schaltplansystem – in das Netzwerk Zelle integriert werden und der evolutionären Veränderung unterliegen. Über unvermutete und neue Wechselwirkungen könnten bei künstlichen biologischen Systemen unerwartete Eigenschaften auftreten und zu unkalkulierbaren Risiken bei einer absichtlichen oder unabsichtlichen Freisetzung von solchen Systemen führen.[72,73,74,75]

Die gleiche Diskussion um die Komplexität biologischer Systeme und potenzieller Risiken gab es Mitte der 1970er-Jahre, nachdem erstmals DNA über Artgrenzen hinweg von einem Organismus auf einen anderen übertragen wurde.[76] Als wesentliche Risiken bei der Herstellung von gentechnisch veränderten Organismen wurden deren absichtliche und unabsichtliche Freisetzung mit unvorhersehbaren Folgen für die Gesundheit von Menschen und Tieren sowie für die Umwelt in ihrem Wirkungsgefüge betrachtet. Diesen Bedenken wurde und wird Rechnung getragen, indem für die Gentechnik ein Risikomanagement etabliert wurde, das für gentechnische Experimente das vermutete Risiko als vorhandenes Risiko annimmt (Vorsorgeprinzip).[77] Mit dem Arbeiten in risikobezogenen Sicherheitslaboren und durch den schrittweisen Übergang vom Sicherheitslabor über zum Beispiel ein Gewächshaus bis hin zur Freisetzung wurde ein technisches Management des angenommenen Risikos von GVO möglich. Mithilfe der biologischen Sicherheitsforschung wurden als biologische Sicherheitsmaßnahmen bezeichnete Vektor-Empfänger-Systeme entwickelt, die außerhalb einer gentechnischen Anlage nicht vermehrungsfähig sind, eine begrenzte Lebenserwartung haben und in einem geringeren Umfang als Wild-

72   Bhutkar A; Synthetic biology: navigating the challenges ahead. J. Biolaw Bus., 2005, 8, 19–29.

73   Church G; Let us go forth and safely multiply. Nature, 2005, 438, 423.

74   Schmidt, M; SYNBIOSAFE – safety and ethical aspects of synthetic biology. 2007, Internet Communication.

75   Tucker JB, Zilinskas RA; The Promise and Perils of Synthetic Biology. The new Atlantis. Spring 2006, 25–45.

76   Berg P, Baltimore D, Boyer HW, Cohen SN, Davis RW, Hogness DS, Nathans D, Roblin R, Watson JD, Weissman S, Zinder ND; Letter: Potential biohazards of recombinant DNA molecules. Science, 1974, 185, 303.

77   Berg P, Baltimore D, Brenner S, Roblin RO (III), Singer MF; Asilomar conference on recombinant DNA molecules. Science, 1975, 188, 991–994.

typorganismen am horizontalen Gentransfer teilnehmen.[78] Dieses Risikomanagement für gentechnische Arbeiten und die beschriebenen Werkzeuge bilden die Grundlage der Risikobewertung für GVO und gentechnische Arbeiten nach dem deutschen Gentechnikgesetz (GenTG), welches die Systemrichtlinie 98/81/EWG und die Freisetzungsrichtlinie 2001/18/EG der EU umsetzt.[79]

Nach dem GenTG entsprechen die meisten der in Kapitel 3 beschriebenen Arbeiten der Synthetischen Biologie gentechnischen Arbeiten. Eine neue Qualität der aus der Gentechnik bekannten Risiken aufgrund des großen Umfangs an neu rekombinierter Nukleinsäuresequenz ist in diesen Arbeiten nicht zu erkennen; in der Gentechnik werden schon seit vielen Jahren Nukleinsäureabschnitte von 50 kb bis mehrere 100 kb über spezielle Vektoren, wie BACs oder YACs, in Zellen übertragen.[80]

Für absichtliche Freisetzungen von Organismen der Synthetischen Biologie, für die kein charakterisierter Referenzorganismus in der Natur existiert, ist vor der Genehmigung einer Freisetzung in die Umwelt die Etablierung neuer Evaluationssysteme (Modellökosysteme wie Mikro- und Mesokosmen) zur Risikoabschätzung zu erwägen. Hier bietet das GenTG die Grundlagen für die Charakterisierung dieser Organismen, damit eine sinnvolle Risikobeurteilung durchgeführt werden kann.

Einige Teilbereiche der Synthetischen Biologie fallen nicht zwangsläufig unter das GenTG. So sind beispielsweise die *de-novo*-DNA-Synthese als Technik der Veränderung genetischen Materials und die Bewertung von mittels Synthetischer Biologie hergestellten Organismen mit einer natürlich vorkommenden Sequenz, die nicht über Rekombinationstechniken zusammengefügt wurde, noch nicht abschließend bewertet. Allerdings ist eine Risikobeurteilung und -kontrolle dieser Organismen mit den Werkzeugen des GenTG problemlos möglich. Eine eventuell in Zukunft notwendige Präzisierung der Zuordnung von Organismen, die nicht von natürlichen Organismen abgeleitet, sondern *de novo* erschaffen werden, sollte bei einer späteren Aktualisierung des GenTG überprüft werden. Das GenTG ist derzeit nicht anwendbar auf artifizielle Zellen, also solche, die nicht fähig sind, sich zu vermehren oder genetisches Material zu übertragen. Aber auch solche Bereiche der Synthetischen Biologie sind über das Chemikaliengesetz, das Arbeitsschutzgesetz und – wenn es sich um Arzneimittel handelt – das Arzneimittelgesetz in eine Risikobewertung zum Schutz

---

78   Kruczek I, Buhk HJ; Risk evaluation. Methods Find Exp. Clin. Pharmacol., 1994, 16, 519–523.

79   Gentechnikgesetz in der Fassung der Bekanntmachung vom 16. Dezember 1993 (BGBl. I S. 2066), zuletzt geändert durch Artikel 1 des Gesetzes vom 1. April 2008. Bundesgesetzblatt, 499.

80   Burke DT, Carle GF, Olson MV; Cloning large segments of exogenous DNA into yeast by means of artificial chromosome vectors. Science, 1987, 263, 806–812.

von Mensch und Umwelt einbezogen.[81,82,83] Aus Sicht der biologischen Sicherheit besitzen weder zellähnliche Systeme noch subgenomische, replikationsdefekte Nukleinsäuren ein Gefährdungspotenzial, weil beide nicht infektiös und nicht vermehrungsfähig sind und sich demzufolge nicht ausbreiten können.

Insofern sind die derzeitigen Arbeiten der Synthetischen Biologie in eine umfassende und ihrem Risiko angemessene Beurteilung eingebunden, sodass augenblicklich keine neuen gesetzlichen Regelungen für erforderlich gehalten werden.

4.3.2 Synthetische Biologie als Sicherheitstechnik

Die in Abschnitt 3.1 dargestellte *de-novo*-Synthese von Nukleinsäuren bietet Möglichkeiten, einen Beitrag zur Erhöhung der Sicherheit bei absichtlicher und unabsichtlicher Freisetzung zu leisten. Vor der Herstellung einer synthetischen Nukleinsäure aus chemischen Bausteinen muss die Sequenzabfolge am Computer definiert werden. Synthetisch hergestellte Elemente oder Organismen besitzen somit eine bekannte Nukleinsäuresequenz. Die Optimierung der *in-vitro*-Synthese von Nukleinsäuren zur Produktion immer längerer Sequenzen ist eine Methode, selten vorkommende Klonierungsartefakte weiter zu minimieren; sie kann darüber hinaus zur Vermeidung von mobilen genetischen Elementen in synthetisch hergestellten Genomen genutzt werden. Durch die *in-vitro*-DNA-Synthese können auch nicht natürliche Nukleotide zur Herstellung der Bauteile und Organismen verwendet werden, die nur von spezifisch veränderten und in der Natur nicht vorkommenden Polymerasen erkannt werden.

Unabhängig von der *in-vitro*-DNA-Synthese ist auch die Verwendung von nicht natürlichen Aminosäuren denkbar, die nur von entsprechend angepassten Ribosomen in Polypeptide eingebaut werden können. Durch die Abhängigkeit von künstlichen Nährstoffen sind die synthetischen Elemente in der Natur nicht aktiv bzw. synthetisch hergestellte Organismen nicht überlebensfähig. Mit der zusätzlichen Integration von synthetischen Schaltkreisen (s. Kap. 3.5) oder Inaktivierungsmechanismen in die Genome von synthetisch hergestellten Organismen und durch die Verwendung nicht natürlicher Nährstoffe ist eine mehrfache Absicherung realisierbar. Die Synthetische Biologie baut somit auf dem Konzept der biologischen Sicherheitsmaßnahmen aus der Gentechnik auf und macht die Minimalzelle, die nur in einer definierten Umgebung eine begrenzte Aufgabe erfüllen kann, zum Ziel einer konsequenten Weiterentwicklung, die das Gefährdungspotenzial im Falle einer Freisetzung weiter verringert.

---

81 Chemikaliengesetz in der Fassung der Bekanntmachung vom 2. Juli 2008. Bundesgesetzblatt, 1146.

82 Arzneimittelgesetz in der Fassung der Bekanntmachung vom 12. Dezember 2005 (BGBl I 3394), zuletzt geändert durch Artikel 9 Abs. 1 des Gesetzes vom 23. November 2007. Bundesgesetzblatt, 2631.

83 Arbeitsschutzgesetz vom 7. August 1996 (BGBl I S. 1246), zuletzt geändert durch Artikel 15 Abs. 89 des Gesetzes vom 5. Februar 2009. Bundesgesetzblatt, 160.

4.3.3 Schutz vor Missbrauch (Biosecurity)

Der vorsätzliche Missbrauch biologischer Substanzen und Organismen für terroristische Zwecke ist eine latente Bedrohung, welche in vielfältigen Variatio-
nen diskutiert wird und unterschiedlichste Szenarien bereithält (zum Beispiel
Anschlagsszenarien durch Pocken, Ebola, Anthrax, Ricin). Geeignete Maßnahmen zum Schutz vor missbräuchlicher Anwendung sind daher notwendig.

Neue technische Methoden zur Genomsequenzierung und die Bereitstellung
von Genomsequenzen in öffentlichen Datenbanken erleichtern grundsätzlich
den Zugang zu genetischen Daten, auch von pathogenen Organismen und biologischen Toxinen. Dieser leichter werdende Zugriff auf Genomdaten und insbesondere die Möglichkeit, definierte Nukleinsäuresequenzen direkt über das
Internet bei DNA-Synthese-Firmen zu bestellen, werden daher als spezifisches
Gefährdungspotenzial der Synthetischen Biologie diskutiert.[84,85,86,87] In diesem
Zusammenhang ist zu bedenken, dass im Bereich der Viren in den vergangenen
Jahren bereits eine Reihe von Genomen hoch pathogener Erreger synthetisiert
wurde, zu denen u.a. das Poliomyelitis-(Kinderlähmungs-)Virus gehört. Es besteht die Befürchtung, dass Einzelpersonen, terroristische Organisationen oder
Staaten damit die Möglichkeit haben, pathogene Organismen oder Toxine zu
rekonstruieren und für feindliche oder kriegerische Handlungen einzusetzen.
Einen ähnlich bedenklichen Ansatz könnten Personen verfolgen, die wie Com-
puter-Hacker und Computer-Virenkonstrukteure als interessierte Laien Zugang
zu einzelnen synthetischen Elementen oder den notwendigen Ausgangsstoffen
bekommen und in einer unkontrollierten Umgebung synthetische Systeme bis
hin zu Mikroorganismen herstellen.

Aufgrund der Vielzahl von Eigenschaften, die einen Krankheitserreger aus-
zeichnen (zum Beispiel Pathogenität, Infektiosität, Wirtsspezifität), wird weniger davon ausgegangen, dass neue, infektiöse Pathogene synthetisch erschaffen werden könnten, sondern vielmehr davon, dass existierende Erreger
rekonstruiert oder modifiziert werden (s. Kap. 3.1). Aufgrund der hohen technischen und logistischen Anforderungen werden die Möglichkeiten von Einzelpersonen, diese Techniken zu missbrauchen, als gering eingeschätzt.

Wie bei allen *dual-use*-Technologien verfolgt der Schutz vor Missbrauch oder
„Biosecurity" auch bei der Synthetischen Biologie das Ziel, die Eventualität eines
Missbrauchs durch gezielte Maßnahmen so weit wie möglich zu minimieren.
In Deutschland existieren verschiedene gesetzliche Regelungen, die das Miss-
brauchsrisiko der Synthetischen Biologie schon jetzt weitgehend einschränken.

---

84   Bhutkar A; Synthetic biology: navigating the challenges ahead. J. Biolaw. Bus., 2005, 8,
     19–29.

85   Schmidt M; SYNBIOSAFE – safety and ethical aspects of synthetic biology. 2007, Internet
     Communication.

86   Schmidt M; Diffusion of synthetic biology: a challenge to biosafety. Syst. Synth. Biol., 2008, 2,
     1–6.

87   Tucker JB, Zilinskas RA; The promise and perils of synthetic biology. New Atlantis, 2006, 12,
     25–45.

Im Gentechnikgesetz wird die Genehmigung zur Errichtung und für den Betrieb einer gentechnischen Anlage abhängig gemacht von der Zuverlässigkeit des Betreibers und der für die Leitung und Aufsicht verantwortlichen Personen. Außerdem dürfen keine Tatsachen vorliegen, die dem Abkommen zu chemischen und biologischen Waffen und dem Kriegswaffenkontrollgesetz entgegenstehen.[88] Nach dem Gesetz über die Kontrolle von Kriegswaffen ist es in Deutschland verboten, biologische oder chemische Waffen zu entwickeln, herzustellen oder mit ihnen Handel zu treiben. Zudem verzichtet die Bundesrepublik Deutschland auf die Herstellung der in der Kriegswaffenliste aufgeführten biologischen Kampfmittel, zu denen genetisch modifizierte Mikroorganismen oder genetische Elemente, die von den in dieser Liste aufgeführten pathogenen Mikroorganismen abstammen, gehören.[89] Nach dem Außenwirtschaftsgesetz bedarf die Ausfuhr von genetischen Elementen und genetisch modifizierten Organismen in Nicht-EU-Staaten einer Genehmigung durch das Bundesamt für Wirtschaft und Ausfuhrkontrolle (BAFA).[90] Einer besonderen Kontrolle unterliegt auch der Versand größerer DNA-Fragmente durch die Gewerbeaufsicht, das BAFA, durch die HADEX und K-Liste, die in besonderem Maße den Versand von Genen oder Genfragmenten einschränkt, die zur Herstellung biologischer Waffen verwendet werden können.

Diese Regularien werden durch freiwillige Selbstverpflichtungen aus Forschung und Industrie zusätzlich unterstützt. Die Deutsche Forschungsgemeinschaft versucht mit dem im April 2008 veröffentlichten Verhaltenskodex[91] für die Arbeit mit hoch pathogenen Mikroorganismen und Toxinen die Aufmerksamkeit insbesondere von Wissenschaftlerinnen und Wissenschaftlern für die Frage des möglichen Missbrauchs von Arbeiten in diesem Gebiet zu wecken und Hinweise für den Umgang zu geben.

In der International Association Synthetic Biology (IASB)[92] oder dem International Consortium for Polynucleotide Synthesis (ICPS)[93] organisierte Unternehmen haben sich in ihren Arbeitsgrundsätzen verpflichtet, die Adressen ihrer Kunden und die zu synthetisierenden Sequenzen auf Pathogenitätsfaktoren und Toxine zu überprüfen und auffällige oder suspekte Aufträge abzulehnen. Die Unternehmen verfolgen teilweise einen sehr konservativen Kurs, indem sie nach der Überprüfung Auftragssynthesen sogar ablehnen, auch, um eine mögliche Gefährdung ihrer eigenen Mitarbeiterinnen und Mitarbeiter auszuschlie-

---

88 Gentechnikgesetz in der Fassung der Bekanntmachung vom 16. Dezember 1993 (BGBl. I S. 2066), zuletzt geändert durch Artikel 1 des Gesetzes vom 1. April 2008. Bundesgesetzblatt, 499. 1-4-2008.

89 Gesetz über die Kontrolle von Kriegswaffen in der Fassung der Bekanntmachung vom 22. November 1990 (BGBl. I S. 2506), zuletzt geändert durch Artikel 24 der Verordnung vom 31. Oktober 2006. Bundesgesetzblatt, 2407. 2006.

90 Außenwirtschaftsgesetz in der Fassung der Bekanntmachung vom 26. Juni 2006. Bundesgesetzblatt, 1386. 2006.

91 www.dfg.de/aktuelles_presse/reden_stellungnahmen/2008/download/codex_dualuse_0804.pdf

92 www.ia-sb.eu/

93 http://pgen.us/ICPS.htm

ßen. Über all diese Maßnahmen hinaus wäre eine Optimierung und Standardisierung der verwendeten Screening-Methoden, mit denen DNA-Sequenzen auf mögliche Pathogenitätsfaktoren oder Toxine untersucht werden, hilfreich. Eine wissenschaftlich fundierte Datenbanklösung zur standardisierten Überprüfung von DNA-Sequenzen erscheint notwendig – diese darf jedoch nicht nur auf Deutschland oder Europa beschränkt bleiben. In Zweifelsfällen brauchen Firmen, die synthetische Nukleinsäuren herstellen, aber eine nationale Kontaktstelle, an die sie sich bei auffälligen Bestellungen wenden können.

Die immer leichtere Verfügbarkeit von DNA-Sequenzen wird zu einer Verbreitung von Techniken der Molekularbiologie und Genetik in andere wissenschaftliche Disziplinen wie zum Beispiel die Ingenieurwissenschaften führen, in denen bisher kaum Erfahrungen im Umgang mit biologischen Agenzien vorliegen. In diesen Bereichen sollte zukünftig die Gentechniksicherheitsverordnung für die Projektleitung gelten.

4.3.4 Begleitendes Monitoring

Die schnellen und vielfältigen Entwicklungen der Synthetischen Biologie lassen nur schwer abschätzen, ob sie zukünftig andere Regelungen verlangen. Daher ist eine kontinuierliche wissenschaftliche Begleitung und gegebenenfalls Evaluation von Fragen der biologischen Sicherheit erforderlich. Der Gesetzgeber sollte die ZKBS mit der sicherheitsrelevanten wissenschaftlichen Begleitung der Synthetischen Biologie beauftragen. Dieses im GenTG verankerte Gremium berät seit 1978 die Bundesregierung und die Länder in Fragen der Sicherheit in der Gentechnik. In der ZKBS wirken neben der Wissenschaft weitere gesellschaftliche Gruppierungen mit, die zum Beispiel den Arbeitsschutz, Verbraucherschutz und Umweltschutz vertreten. In Kooperation mit den für die Genehmigung und Überwachung von gentechnischen Arbeiten und Anlagen zuständigen Behörden der Länder und des Bundes hat sie in den letzten 30 Jahren ein gesellschaftlich akzeptiertes System der Risikobewertung für im Genom modifizierte Organismen mit entwickelt. In diesem System wird kontinuierlich der aktuelle Stand von Wissenschaft und Technik berücksichtigt. Anhand ihrer Sach- und Fachkompetenz ist die ZKBS in der Lage, die wissenschaftliche Literatur zur Synthetischen Biologie sicherheitsbezogen zu verfolgen. Darüber hinaus könnte die oben vorgeschlagene Kontaktstelle in Kooperation mit der ZKBS bei Erkennbarwerden neuer Risiken Ansatzpunkte zur Justierung der bestehenden Regularien an die Anforderungen des Gefährdungspotenzials der Synthetischen Biologie erarbeiten.

Wie schon bei der Gentechnik sollten eventuell notwendige, dann noch auszuarbeitende Regeln für die Überwachung und Kontrolle der Forschung und Anwendung der Synthetischen Biologie nicht nur von einzelnen Staaten national aufgestellt werden, sondern als international anerkannte Grundsätze formuliert werden, an denen sich nationale Regelungen orientieren.

4.4 Ethische Fragen

Die beispielsweise im Abschnitt 4.3 diskutierten Fragen nach unbeabsichtigten Schäden oder vorsätzlichem Missbrauch im Zusammenhang mit der Synthetischen Biologie sind für deren ethische Beurteilung ebenso relevant wie die in Abschnitt 4.1 implizierten Gerechtigkeitsfragen etwa im Zusammenhang mit geistigen Eigentumsrechten, Patenten und Nutzungsrechten. Solche Probleme sind – was sie keineswegs relativiert – im Prinzip aus anderen Sektoren der modernen biomedizinischen Forschung bekannt und sollten vor diesem Hintergrund diskutiert und gehandhabt werden.

Da weite Bereiche der Synthetischen Biologie eine Weiterentwicklung der molekularen Biologie und Gentechnik darstellen, sind viele bewährte Methoden der Technikfolgenabschätzung und der Risikobeurteilung anwendbar. Allerdings sind in den Fällen, bei denen es keine natürlichen Referenzsysteme gibt, neue Maßstäbe für die Risikobeurteilung notwendig. Denn mit der Neuentwicklung von synthetischen Organismen eröffnen sich noch wenig erforschte Unsicherheitsspielräume, die einen sorgfältigen Umgang erforderlich machen. Vor allem bei hoher Komplexität und Unsicherheit sind die Regeln des Vorsorgeprinzips anzuwenden. Darunter fallen vor allem das Prinzip des „containment" von Anwendungen (räumliche oder zeitliche Begrenzung), ein intensives Monitoring der Folgen und eine flexible problemgerechte Anpassung der Regulierung an die empirische Praxis. Für die Beurteilung der Folgen sind Szenarien zu erarbeiten, die auch unbeabsichtigte Schädigungen von Menschen, Landwirtschaft und Umwelt berücksichtigen. Manche Risiken können durch spezifische Mechanismen der Synthetischen Biologie verringert werden, etwa dadurch, dass die hergestellten Entitäten außerhalb des Labors voraussichtlich nicht überlebensfähig sind oder nicht an der Evolution teilhaben. Grundsätzlich ist aus ethischer Sicht der mögliche Schaden (Risiko) gegen den möglichen Nutzen (Chancen) abzuwägen.

Genuin neue ethische Fragen sehen manche Bioethiker durch den Anspruch der Synthetischen Biologie aufgeworfen, neuartiges Leben zu erschaffen. Hier nämlich gehe es um fundamentale und neue Aspekte unseres Verständnisses von Leben im Gegensatz zu Artefakten oder Maschinen, um Fragen nach Wert und Gefährdung des Lebendigen und insofern auch um das Selbstverständnis des Menschen.[94] Schon diese These von der Neuartigkeit der ethischen Fragen wird allerdings von anderen Bioethikern bestritten, die keinen Bedarf für eine eigene „Synthetic Bioethics" sehen,[95] sondern die genannten Fragen als Facetten bekannter Probleme ansehen und behandeln wollen – bekannt aus den Debatten zur Herstellung transgener Pflanzen und Tiere, zum Klonen, zur

---

94 So Boldt J, Müller O; Newtons of the leaves of grass. Nat. Biotechnol., 2008, 26, 387–389; Boldt J, Müller O, Maio G; Synthetische Biologie. Eine ethisch-philosophische Analyse. 2009, Bern: Kap. 6.

95 Zum Beispiel: Parens E, Johnston J, Moses J; Ethics. Do we need "synthetic bioethics"? Science, 2008, 321, 1449.

Chimärenbildung oder Zellreprogrammierung, aber auch zur assistierten Reproduktion und zum genetischen Enhancement.

Unstrittig ist jedoch, dass diese Fragen unter den Experten für Ethik aufgearbeitet und dann in die öffentliche Diskussion eingebracht werden sollen. Dies sollte bereits im Vorfeld der geplanten technischen Weiterentwicklungen geschehen. Es ist vorstellbar, dass für die strukturierte Diskussion eine entsprechende Plattform vorgesehen wird.

Für diese anstehenden Debatten lassen sich einige Thesen und Desiderate formulieren.

(1) Es ist weder das Ziel noch ein für absehbare Zeit realistisch erscheinendes Ergebnis der Synthetischen Biologie, durch Synthese oder Manipulation neuartige höhere Lebewesen zu schaffen. Es geht ihr vielmehr um die Veränderung und die *de-novo*-Synthese von Mikroorganismen, einzelnen Zellen und Zellpopulationen. Gleichwohl führt bereits diese begrenzte Zielsetzung zu grundlegenden Fragen nach der Definition des Lebens; auch sollten weitergehende Optionen zumindest hypothetisch im Auge behalten werden.

(2) Unser alltägliches Vorverständnis von ‚Leben' wird von einer Pluralität zum Teil unvereinbarer kultur- und traditionsrelativer Kriterien bestimmt (morphologische Schemata, religiös geprägtes Naturverständnis, naturwissenschaftliche Allgemeinbildung). Darüber hinaus gehen aber auch verschiedene wissenschaftliche Disziplinen mit ihren spezifischen Forschungsansätzen und Zielen von einem unterschiedlichen Verständnis des Lebens aus. Wenn man zum Beispiel ein in den Naturwissenschaften gängiges Konzept zur Definition des Lebens verallgemeinern würde, wonach die Aufrechterhaltung des Stoffwechsels, die Fähigkeit zur evolutionären Veränderung und die Fähigkeit zur Reproduktion drei notwendige Bedingungen von Leben sind, würden etwa Maultiere, die wie viele Hybride[96] nicht fortpflanzungsfähig sind, nicht unter die Definition des Lebendigen (und damit zum Beispiel auch nicht unter die Tierschutzgesetze) fallen – ein offensichtlich unangemessenes Ergebnis. Für eine effiziente, in verständlicher und verlässlicher Kommunikation geführte Debatte über die Herausforderungen der Synthetischen Biologie bedarf es deshalb einer problem-angemessenen, möglichst einheitlichen Bestimmung des Lebendigen und einer möglichst eindeutigen Abgrenzung gegen das Nichtlebendige. Von daher sind die von manchen Vertreterinnen und Vertretern der Synthetischen Biologie verwendeten Begriffe und Metaphern (zum Beispiel ‚lebendige Maschinen') semantisch problematisch, indem sie die Grenze zwischen Lebendigem und ‚toter Materie' zu verwischen scheinen.

(3) Bei der Beschreibung von Entitäten ist bereits begrifflich – und *vor* aller Bewertung – zwischen ihren Eigenschaften, etwa ihren Funktionsfähigkeiten und Entwicklungspotenzialen, und den Bedingungen ihrer Entstehung (durch natürliche Prozesse, durch Synthese oder durch genetische Eingriffe) zu unter-

---

96   Hier ist das Hybrid aus einer Kreuzung zwischen einer Pferdestute und einem Eselshengst hervorgegangen.

scheiden. Nur so lässt sich der potenziellen Komplexität denkbarer Formen des Lebendigen gerecht werden.

(4) Moralische Argumente zugunsten der Herstellung synthetischen Lebens beziehen sich auf den erhofften Nutzen für Medizin, Landwirtschaft, Energieproduktion oder Umwelt, dem zufolge die Anwendung der Synthetischen Biologie nicht nur erlaubt, sondern sogar geboten ist. Ferner wird die Synthetische Biologie unter Hinweis auf ökonomische Vorteile und schließlich auf die Forschungsfreiheit gerechtfertigt, die allerdings nach allgemeinem Konsens durch andere Grundrechte wie das Recht auf körperliche Unversehrtheit in Schranken gehalten wird.

(5) Zu den *fundamentalen* ethischen Einwänden *gegen* Anwendungen der Synthetischen Biologie könnten gehören:

(a) dass diese unzulässig in die Schöpfung oder sakrosankten Prozesse der Natur eingriffen (man spiele Gott),

(b) dass sie durch die Herstellung neuartiger Lebewesen die Integrität der Natur zerstöre bzw. die Ordnung der Lebewesen und Arten beschädige oder

(c) dass wir das Leben im Zuge seiner fortschreitenden ‚Herstellbarkeit‘ vielleicht nicht mehr in angemessener Weise respektieren und schützen würden.[97]

Die beiden ersten Arten von Einwänden leben von starken weltanschaulichen bzw. metaphysischen Prämissen, die sicher nicht von allen Menschen, auch innerhalb von religiösen Gemeinschaften, geteilt werden.

(a) Argumenten des unzulässigen Eingriffs in die Schöpfung oder in die Abläufe der Natur liegt etwa die religiöse Vorstellung zugrunde, nur Gott dürfe Leben schaffen. Hier werden also nicht die möglichen Produkte der Eingriffe kritisiert, sondern der Prozess ihrer Herstellung. Doch auch wenn man zugesteht, dass die Welt von einem Gott erschaffen wurde, folgt daraus noch nicht, dass es dem Menschen verboten sein soll, Leben synthetisch zu erzeugen. Wenn man unterstellt, dass allen oder einigen Lebewesen ein eigenständiger intrinsischer Wert zukommt, ist zudem keineswegs ausgeschlossen, dies auch auf synthetisch hergestelltes Leben zu beziehen. Und schließlich lässt sich nicht plausibel machen, warum andere tief gehende Eingriffe in die Natur (zum Beispiel medizinische Behandlungen) dann grundsätzlich positiver beurteilt werden dürften.

(b) Auch Argumente, denen zufolge es ethisch problematisch ist, *neuartige*, also in der bisherigen Natur nicht vorkommende Lebewesen herzustellen, können in dieser Grundsätzlichkeit nicht überzeugen. So lässt sich die Vorstellung von einer an sich integren Natur, die lediglich durch den Menschen gestört wird, kaum mit elementaren Erfahrungen von Selbstzerstörung der Natur, natürlicher Aggression, dem Vorkommen von Seuchen und schweren Krankheiten usw. in Einklang bringen. Überdies widerspricht die Idee einer fixen und sakrosankten Ordnung der Lebewesen und Arten bereits den na-

---

97 Vgl. Boldt J, Müller O, Maio G; Synthetische Biologie. Eine ethisch-philosophische Analyse. 2009, Bern: Kap. 6.

türlichen biologischen Phänomenen der Veränderung, der Durchmischung oder dem Aussterben von Arten.

(c) Der Einwand, die Anwendungen der Synthetischen Biologie könnten unser Grundverständnis vom Leben im Allgemeinen und von der Schutzwürdigkeit menschlichen Lebens im Besonderen negativ beeinflussen, bedarf in seiner Bedrohlichkeit gewiss der gründlichen Analyse, erscheint aber doch auf den ersten Blick einigermaßen spekulativ. Die Baupläne des Lebendigen besser verstehen, reproduzieren oder manipulieren zu können, sollte an unseren ethischen Einstellungen gegenüber Natur und Individuen ebenso wenig etwas ändern, wie es die teilweise Beherrschbarkeit krankhafter Veränderungen getan hat.

(6) Die Debatte über die Selbstregulierung der Wissenschaft wird gerade in Bezug auf die Synthetische Biologie kontrovers geführt. In der Wissenschaft wird die verantwortliche Wahrnehmung der Forschungsfreiheit durchaus ernst genommen. Im Jahr 2006 wurden auf der Tagung „SyntheticBiology 2.0" in Berkeley Konzepte zur Selbstregulierung diskutiert und der Öffentlichkeit vorgestellt[98]. Dabei wurden vor allem Wege gesucht, eine Balance zwischen freier Zugänglichkeit von Daten und der Verhinderung von deren Missbrauch zu finden. In einem offenen Brief haben allerdings 35 NGOs diesen Ansatz der Selbstregulierung als nicht ausreichend kritisiert und einen weiterreichenden gesellschaftlichen Dialog gefordert.[99] In dem Brief wird eine Parallele zu der „Asilomar Conference on Recombinant DNA" gezogen, bei der 1975 eine Gruppe von 140 Wissenschaftlerinnen und Wissenschaftlern auf der Basis des Vorsorgeprinzips zu Selbstregulierung im Umgang mit rekombinanter DNA aufgerufen hatte. Dieser Aufruf habe dazu geführt, dass die Kontrolle der Gentechnologie lange Zeit in zu großem Maße der Wissenschaft überlassen wurde. Die Ansätze der Selbstregulierung der Wissenschaft sind seit dem oben genannten Konzeptvorschlag bisher nicht fortgesetzt worden.

(7) Alle diese Überlegungen gilt es für das innovative Forschungsgebiet der Synthetischen Biologie gründlich, interdisziplinär und kontextübergreifend zu diskutieren. Erforderlich sind daher eine frühzeitige ethische Begleitforschung und kritische Reflexion auf die verantwortungsvolle Wahrnehmung der Forschungsfreiheit in der Wissenschaft. Zudem bedarf es intensiver Bemühungen, die Öffentlichkeit frühzeitig über das Geschehen im Labor aufzuklären, Risiken und Chancen aufzuzeigen und die ethische Reflexion zu ermöglichen.[100]

---

98   Vgl. Schmidt M, Torgersen H, Ganguli-Mitra A, Kelle A, Deplazes A, Biller-Andorno N; „SYN-BIOSAFE e-conference: online community discussion on the societal aspects of synthetic biology", in: Systems and Synthetic Biology (Online First Publication, 2008 Sep 18): 11 S. Public declaration from the Second International Meeting on Synthetic Biology (May 20–22, 2006, Berkeley, CA); http://hdl.handle.net/1721.1/32982

99   NEWS RELEASE, 19th May 2006, Global Coalition Sounds the Alarm on Synthetic Biology, Demands Oversight and Societal Debate; www.etcgroup.org/en/materials/publications.html?pub_id=8

100 So auch Schmidt M, Torgersen H, Ganguli-Mitra A, Kelle A, Deplazes A, Biller-Andorno N; "SYNBIOSAFE e-conference: online community discussion on the societal aspects of synthetic biology". In: Systems and Synthetic Biology (Online First Publication, 2008 Sep 18): 11 S.

Anhang

A) Textgenese und Zusammensetzung der Arbeitsgruppe

Die drei beteiligten Organisationen (DFG, acatech und Leopoldina) haben zunächst einen gemeinsamen Workshop vorbereitet, der am 27. Februar 2009 in Berlin stattfand (Programm siehe Anhang B). Die Referenten und Teilnehmer legten mit ihren Vorträgen und Diskussionsbeiträgen die Grundlage für diese Stellungnahme, die im Anschluss an den Workshop von einer interdisziplinären Arbeitsgruppe „Synthetische Biologie" unter dem Vorsitz von Frau Prof. Dr. Bärbel Friedrich als Vorsitzende der Senatskommission für Grundsatzfragen der Genforschung der DFG erarbeitet wurde. Die Mitglieder der Arbeitsgruppe sind unten aufgeführt. Für die Erstellung einzelner Textteile wurden weitere Expertinnen und Experten hinzugezogen. Die Stellungnahme wurde einem Begutachtungsprozess unterzogen und anschließend von den Präsidien der DFG, acatech und Leopoldina verabschiedet.

Mitglieder der Arbeitsgruppe

Professor Dr. Christopher Baum Mitglied der Senatskommission für Grundsatzfragen der Genforschung	Medizinische Hochschule Hannover Abteilung Experimentelle Hämatologie Carl-Neuberg-Straße 1 / OE 6960, K 11, Raum 1120 30625 Hannover
Dr. Matthias Brigulla	Bundesamt für Verbraucherschutz Referat 402 Mauerstraße 39–42 10117 Berlin (zurzeit abgeordnet an das Bundesministerium für Ernährung, Landwirtschaft und Verbraucherschutz)
Professor Dr. Bärbel Friedrich Mitglied der Senatskommission für Grundsatzfragen der Genforschung Vizepräsidentin der Leopoldina Vorsitzende der Arbeitsgruppe	Humboldt-Universität zu Berlin Institut für Biologie Chausseestraße 117 10115 Berlin

Professor Dr. Carl F. Gethmann Mitglied acatech Mitglied der Leopoldina	Universität Duisburg-Essen Fachbereich Geisteswissenschaften Institut für Philosophie Universitätsstraße 12 45141 Essen
Professor Dr. Jörg Hacker Vizepräsident der DFG Mitglied der Senatskommission für Grundsatz- fragen der Genforschung Mitglied der Leopoldina	Robert Koch-Institut (RKI) Nordufer 20 13353 Berlin
Professor Dr. Klaus-Peter Koller Mitglied der Senatskommission für Grundsatz- fragen der Genforschung	Sanofi-Aventis Deutschland GmbH F&E, External Innovation Bldg. H 831 Industriepark Höchst 65926 Frankfurt
Professor Dr. Bernd Müller-Röber Mitglied acatech	Universität Potsdam Mathematisch-Naturwissenschaftliche Fakultät Institut für Biochemie und Biologie Karl-Liebknecht-Straße 24–25 14476 Golm
Professor Dr. Alfred Pühler Mitglied acatech Mitglied der Leopoldina	Universität Bielefeld Centrum für Biotechnologie (CeBiTec) Universitätsstraße 27 33615 Bielefeld
Professor Dr. Bettina Schöne-Seifert	Universitätsklinikum Münster Institut für Ethik, Geschichte und Theorie der Medizin Von-Esmarch-Straße 62 48149 Münster
Professor Dr. Jochen Taupitz	Universität Mannheim Institut für Deutsches, Europäisches und Internationales Medizinrecht, Gesundheitsrecht und Bioethik der Universitäten Heidelberg und Mannheim Schloss / Postfach 68131 Mannheim
Professor Dr. Rudolf Thauer Mitglied des Präsidiums der Leopoldina	Max-Planck-Institut für terrestrische Mikro- biologie Karl-von-Frisch-Straße 35043 Marburg
Professor Dr. Angelika Vallbracht Mitglied der Senatskommission für Grundsatz- fragen der Genforschung	Universität Bremen Zentrum für Umweltforschung und nachhaltige Technologien (UFT) Abteilung Institut für Virologie Postfach 330440 28359 Bremen

Von den Geschäftsstellen

Dr. Ingrid Ohlert	Deutsche Forschungsgemeinschaft Fachgruppe Lebenswissenschaften Kennedyallee 40 53175 Bonn
Dr. Nikolai Raffler	Deutsche Forschungsgemeinschaft Fachgruppe Lebenswissenschaften Kennedyallee 40 53175 Bonn
Dr. Marc-Denis Weitze	acatech – Deutsche Akademie der Technikwissenschaften Projektzentrum Residenz München Hofgartenstraße 2 80539 München

Für die Unterstützung bei der Ausarbeitung danken wir Herrn Prof. Dr. Nediljko Budisa (Max-Planck-Institut für Biochemie, Planegg), Herrn Dr. Jürgen Eck (B.R.A.I.N. AG, Darmstadt), Frau Dr. Margret Engelhard (Europäische Akademie zur Erforschung von Folgen wissenschaftlich-technischer Entwicklungen Bad Neuenahr-Ahrweiler GmbH), Herrn Prof. Dr. Jürgen Heesemann (Universität München), Herrn Prof. Dr. Hans-Dieter Klenk (Universität Marburg) und Frau Prof. Dr. Petra Schwille (Technische Universität Dresden). Wir danken Dr. Robin Fears (EASAC London, UK) für seine redaktionelle Mitarbeit bei der englischen Übersetzung.

Besonderer Dank gilt den Mitgliedern der Senatskommission für Grundsatzfragen der Genforschung, die die Erarbeitung der Stellungnahme in der DFG initiiert hat, für ihre hilfreichen Kommentare und ihre Unterstützung *(www.dfg.de/ dfg_im_profil/struktur/gremien/senat/kommissionen_ausschuesse/senatskommission_ grundsatzfragen_genforschung/index.html)*.

Für die Vorbereitungsphase gilt unser Dank den Sprechern und Teilnehmern des Workshops „Synthetische Biologie" am 27. Februar 2009 in Berlin.

B) Programm des Workshops

Workshop „Synthetic Biology"

Thursday, 26 February 2009 – Hotel NH Berlin Mitte (Leipziger Straße 106–111)

Arrival of participants	
19.30 – 21.00	Reception at the Hotel (Prof. Matthias Kleiner)

Friday, 27 February 2009 – Landesvertretung Sachsen-Anhalt (Luisenstraße 18)

09.00 – 09.20	Welcome (Prof. Matthias Kleiner, Prof. Reinhard Hüttl, Prof. Bärbel Friedrich)
Part I	**Moderation: Prof. Jörg Hacker**
09.20 – 09.55	Minimal Genomes (Prof. György Pósfai, Szeged, HU)
09.55 – 10.30	Protocells (Prof. John McCaskill, Bochum, DE)
10.30 – 11.05	Orthogonal Biosystems (Prof. Jason Chin, Cambridge, UK)
11.05 – 11.25	Coffee Break
11.25 – 12.00	Genetic Circuits (Prof. Martin Fussenegger, Zürich, CH)
12.00 – 13.30	Discussion (Prof. Alfred Pühler)
13.30 – 14.30	Lunch Break
Part II	**Moderation: Prof. Bärbel Friedrich**
14.30 – 15.05	Ethical Issues (Prof. Paul Martin, Nottingham, UK)
15.05 – 15.40	Socioeconomical Issues (Prof. Ralf Wagner, GeneArt, Regensburg, DE)
15.40 – 16.15	Legal Issues (Dr. Berthold Rutz, European Patent Office, München, DE)
16.15 – 16.45	Coffee Break
16.45 – 17.20	Biosafety and Biosecurity Issues (Dr. Markus Schmidt, Vienna, AT)
17.20 – 18.50	Discussion (Prof. Klaus-Peter Koller)
18.50 – 19.00	Closing Remarks (Prof. Rudolf Thauer)
19.00 – 21.00	Dinner – Restaurant „Habel Weinkultur" (Luisenstraße 19)

Saturday, 28 February 2009 – Hotel NH Berlin Mitte (Leipziger Straße 106–111)

Departure	

C) Glossar

BAC: Bacterial Artificial Chromosome; Vektor zur → Klonierung von großen Genomabschnitten in Bakterien, zum Beispiel *Escherichia coli.*

BAFA: Bundesamt für Wirtschaft und Ausfuhrkontrolle

BioBrick: Charakterisierter genetischer Baustein oder genetisches Schaltelement.

BMBF: Bundesministerium für Bildung und Forschung

bp: Basenpaar

cDNA: (engl. *complementary DNA*). Es handelt sich um eine → DNA, die i.d.R. mittels des Enzyms reverse Transkriptase meist aus → mRNA synthetisiert wird.

Codon: → Kodon

de novo: (lat.) von Neuem, von Grund auf

DNA: Desoxyribonukleinsäure (engl. *desoxyribonucleic acid*, DNA); chemischer Grundbaustein der Erbsubstanz. Die DNA enthält die Informationen für die Herstellung aller für die Körperfunktionen nötigen Eiweiße.

EPÜ: Europäisches Patentübereinkommen

Expression: → Genexpression ist das Umsetzen der Information, die in der DNA eines Gens gespeichert ist, zu Zellstrukturen und Signalen. Diese liegen oft in Form von Proteinen vor. Die Expression von Genen ist ein komplexer Prozess, der aus vielen verschiedenen Einzelschritten besteht. Generell kann die Regulation der Genexpression auf verschiedenen Stufen des Realisierungsprozesses vom Gen zum Merkmal führen.

Gen: DNA-Abschnitt, der für eine Funktion, beispielsweise ein Protein, kodiert. Neben den kodierenden Bereichen (Exons) umfassen Gene weitere Regionen wie Introns (nicht kodierende Abschnitte) und → Promotoren (Regulationselemente).

Genexpression: Umsetzung der genetischen Information, meist in Form von Proteinen, zur Bildung von Zellstrukturen und Signalen.

Genom: Nicht einheitlich gebrauchter Begriff für die Gesamtheit der → DNA eines Individuums oder der genetischen Information einer Zelle (→ Gen).

Gentechnik: Biotechnologische Methoden und Verfahren der Biotechnologie, die gezielte Eingriffe in das Erbgut (→ Genom) und damit in die biochemischen Steuerungsvorgänge von Lebewesen bzw. viralen Genomen ermöglichen.

GenTG: Gentechnikgesetz

Gentherapie: → Somatische Gentherapie

Gentransfer: Der methodische Vorgang des Einbringens von Genen in Zellen.

GenTSV: Gentechnik-Sicherheitsverordnung

GVO: Gentechnisch veränderter Organismus. Organismus, dessen Erbanlagen mittels gentechnischer Methoden gezielt verändert wurde.

HADEX-Liste: Eine Ausschlussliste, die Kunden (Firmen, Einrichtungen) aufführt, die keine *dual-use*-Güter erhalten dürfen.

IASB: International Association Synthetic Biology

ICPS: International Consortium for Polynucleotide Synthesis

in silico: (angelehnt an lat. *in silicio,* in Silicium); Vorgänge, die im Computer ablaufen.

in vitro: (lat.) im Glas (Reagenzglas, in Zellkultur etc.); gemeint ist die Erzeugung außerhalb des Organismus, im Unterschied zu → *in vivo,* im lebenden Organismus.

in vivo: (lat.) im Lebendigen; Prozesse, die im lebenden Organismus ablaufen.

Insertion: Einschub, Einbau; hier: Einbau von DNA-Sequenzen in ein Genom.

kb: Kilobasen = 1000 Basen

K-Liste: Eine Ausschlussliste, die Staaten aufführt, die keine *dual-use*-Güter erhalten dürfen.

Klonen: → Klonierung

Klonierung: Man versteht darunter das Kopieren und identische Vermehren. Der Begriff wird im Zusammenhang mit Molekülen, Zellen, Geweben, Pflanzen (Ableger), Tieren und Menschen verwendet. Klone werden als gen-identische Kopien bezeichnet.

Kodon: Bezeichnung für eine Sequenz von drei → Nukleobasen (Basentriplett) der → mRNA, die im genetischen Code für eine Aminosäure kodiert.

Mb: Megabasen = 1 000 000 Basen

Metabolom: Gesamtheit aller Metabolite

mRNA: (engl. *messenger RNA*) Boten-RNA; Bezeichnung für das → Transkript eines zu einem → Gen gehörenden Teilabschnitts der DNA.

NGO: (engl. *non governmental organisation*); Nicht-Regierungs-Organisation

Nukleinsäure: Aus einzelnen Bausteinen, den → Nukleotiden, aufgebaute Makromoleküle. Siehe auch → DNA.

Nukleobase: → Nukleotid

Nukleotid: Grundbaustein von → Nukleinsäuren (→ DNA und → RNA).

Oligonukleotid: (griech. *oligo,* wenige); aus wenigen → Nukleotiden (→ DNA oder → RNA) aufgebaute Oligomere.

PatG: Patentgesetz

Plasmid: Kleine, in der Regel zirkuläre, autonom replizierende DNA-Moleküle, die in Bakterien extrachromosomal vorkommen, sie können mehrere → Gene enthalten.

Promoter: (ursprünglich franz. *promoteur,* Anstifter, Initiator); Bezeichnung für eine DNA-Sequenz, welche die regulierte → Expression eines → Gens ermöglicht. Die Promotorsequenz ist ein essenzieller Bestandteil eines Gens.

Proteom: Gesamtheit aller → Proteine

Ribosom: (*griech.* Αραβινόζ, *arabinos,* Traube, und σωμα, *soma,* Körper); hoch spezialisierter Komplex, bestehend aus Proteinen und RNA, der einen zentralen Teil der Proteinbiosynthese vermittelt: die in der Sequenzabfolge der → mRNA gespeicherte Information wird abgelesen und in die Herstellung von Proteinen umgesetzt.

RNA: (engl. *ribonucleic acid,* RNA); Ribonukleinsäure; Informationsspeicher auf Nukleinsäurebasis mit wesentlicher Funktion bei der Umsetzung von Erbinformation in Proteine (→ Transkription).

rRNA: ribosomale RNA

Somatische Gentherapie: Anwendung des Gentransfers auf somatische Zellen (→ Somatische Zellen) mit dem Ziel der Prävention oder Behandlung von Erkrankungen. Genetische Veränderungen werden hierbei nicht an die Nachkommen weitergegeben.

Somatische Zellen: Körperzellen, deren genetische Information nicht an nachfolgende Generationen weitervererbt werden kann. Sie bilden den Großteil der menschlichen Zellen, lediglich Keimzellen (Ei- und Samenzellen) können Erbinformationen auf die nächste Generation übertragen und bilden die sogenannte Keimbahn (→ Somatische Gentherapie).

Systembiologie: Zweig der Biowissenschaften, der versucht, biologische Systeme und Prozesse quantitativ in ihrer Gesamtheit zu verstehen.

transgen: Es handelt sich dabei i.d.R. um einen gentechnisch veränderten Organismus (→ GVO), der in seinem → Genom zusätzliche → Gene anderer Arten enthält.

Transkript: → Transkription

Transkription: (lat. *trans*, jenseits, hinüber; *scribere*, schreiben); Transkription ist in der Biologie der erste Schritt der Proteinbiosynthese, der zur Bildung der → mRNA führt; hierzu zählt auch die Synthese der → tRNA und der → rRNA. Bei der Transkription wird ein → Gen abgelesen und als mRNA-Molekül vervielfältigt, das heißt, ein spezifischer DNA-Abschnitt dient als Vorlage zur Synthese eines neuen RNA-Strangs. Bei diesem Vorgang werden die Nukleinbasen der DNA (T, A, G, C) in die Nukleinbasen der RNA (U, A, G, C) umgeschrieben.

Transkriptom: Gesamtheit aller → Transkripte

Transposon: Genabschnitt, der die Möglichkeit hat, seinen Ort innerhalb des → Genoms zu verändern (= Transposition).

tRNA: transfer-RNA

Vakzin: Ein biologisch oder gentechnisch hergestelltes Antigen, meist bestehend aus Protein- oder Erbgutbruchstücken, abgetöteten oder abgeschwächten Erregern. Der Impfstoff dient im Rahmen einer Impfung zur spezifischen Aktivierung des Immunsystems hinsichtlich eines bestimmten Erregers bzw. einer Erregergruppe.

YAC: Yeast Artificial Chromosome; Vektor zur → Klonierung von großen Genomabschnitten in Hefezellen.

ZKBS: Zentrale Kommission für die Biologische Sicherheit

Synthetic Biology

Statement

Contents

Foreword

A new research field known as synthetic biology is emerging from an interdisciplinary combination of biology, molecular biology, chemistry, biotechnology, information technology and the engineering sciences. It has recently become a focus of attention, both nationally and internationally.

Synthetic biology can make a major contribution to gaining knowledge in basic research. Furthermore, it provides medium-term opportunities for biotechnological applications, including new and improved diagnostic agents, vaccines and medicines, and in the development of new biosensors, biomaterials and biofuels.

However, this area of research raises concerns regarding, for example, legal aspects relating to biological safety and protection against misuse, commercial use and ethical aspects.

In view of this, the Deutsche Forschungsgemeinschaft (DFG, German Research Foundation), acatech – Deutsche Akademie der Technikwissenschaften (German Academy of Science and Engineering), and the Deutsche Akademie der Naturforscher Leopoldina – Nationale Akademie der Wissenschaften (German Academy of Sciences Leopoldina) have joined forces to prepare a joint statement on the possible opportunities and risks associated with synthetic biology.

In order to encourage a constructive dialogue between the disciplines, the three organisations held a joint international workshop. Scientists from the fields of biochemistry, molecular biology, genetics, microbiology, virology, chemistry, and physics as well as from the social sciences and humanities met with representatives from public institutions and industry to exchange information. The following statement is based on information from presentations and in-depth discussions. It is directed at representatives of political bodies and the authorities, the public and last but not least, the scientific community.

At present, synthetic biology is mainly concentrated on fundamental research. But like any new technology that may exert a major influence, it is not only the economic opportunities and scientific research agenda that are of interest; it is also important to deal with unintentional consequences at an early stage. This means that risks and opportunities should be evaluated as far as possible and the lessons learned must be incorporated in the design and in the conditions for application of the new technology. Moreover, like any new technology, an early and open dialogue with the public is vital. This is the only way to ensure a responsible climate of innovation in a democratic and pluralistic society.

Thus, the expectations for new knowledge are high. But because the opportunities and challenges require careful consideration, there is also the need for a broadly based scientific and public discussion of the issues relating to possible future applications.

July 2009

Prof. Dr.-Ing. Matthias Kleiner

Präsident
Deutsche
Forschungsgemeinschaft
(German Research
Foundation)

Prof. Dr. Reinhard Hüttl

President
acatech –
Deutsche Akademie der
Technikwissenschaften
(German Academy of
Science and Engineering)

Prof. Dr. Volker ter Meulen

President
Deutsche Akademie der
Naturforscher Leopoldina –
Nationale Akademie
der Wissenschaften
(German Academy of
Sciences Leopoldina)

1 Summary and Recommendations

Synthetic biology is based on knowledge of molecular biology, decoding of the complete genome, the holistic consideration of biological systems and technical advances in the synthesis and analysis of nucleic acids. It combines a wide spectrum of scientific disciplines and follows the principles of engineering science. The chief characteristic of synthetic biology is the modification of biological systems, which may also be combined with chemically synthesised components to produce new entities. This may give rise to properties not previously exhibited by naturally occurring organisms.

The term synthetic biology covers a research and application field that cannot be strictly differentiated from conventional genetic engineering and biotechnological processes. It can therefore be regarded as a further development of these disciplines and their respective objectives. The first part of this statement deals with selected areas of synthetic biology that are orientated toward basic research:

► Technological advances in the synthesis and analysis of nucleic acids. These facilitate not only recombinant genetic engineering processes, but also enable major advances in gene therapy.
► Construction of minimal cells with synthetically produced or genetically downsized genomes in order to produce the smallest viable unit. Although such cells are viable under defined laboratory conditions, their ability to reproduce in natural habitats is limited.
► Synthesis of protocells with characteristics of living cells. The long-term aim – as for minimal cells – is to use them as a "chassis" for the biosynthesis of substances.
► Production of new biomolecules by modular assembly of individual metabolic functions. These may originate from a wide range of donor organisms.
► Construction of regulatory circuits that respond to external stimuli. These can be used to control complex biological or synthetic processes.
► Design of so-called "orthogonal systems". This involves the use of modified cellular machineries, for example, to create novel biopolymers.

Most of the current work in the field of synthetic biology is still at the basic research level. It can be expected to provide important scientific information that will have a long-term effect on the development of new pharmaceuticals and therapies as well as the production of industrial chemicals and the design of catalytic processes. Synthetic biology can be used to produce organisms that are only able to survive under controlled conditions.

But how can the market potential be evaluated? What are the basic scientific conditions? Is synthetic biology associated not only with these diverse opportunities but also with potential risks? These questions are addressed in the second part of this statement. The following aspects are being discussed:

▶ Although the economic implications of synthetic biology cannot be precisely evaluated at present, some market-oriented product concepts are already emerging that offer very promising prospects not only for industrial utilisation but also for society in general. The catalogue includes drugs, nucleic acid vaccines, novel gene therapies, environmentally friendly and resource-conserving fine and industrial chemicals, biofuels and new materials such as polymeric compounds.

▶ The scientific conditions for synthetic biology in Germany are regarded as favourable. The first funding programmes that are dedicated to these disciplines have already been initiated on the European and national levels. Owing to the overlap with conventional biotechnological and biomolecular projects, synthetic biology projects are also being funded via other scientific areas. The basic infrastructures are available or can easily be established and extended in existing research centres. The traditional strength in the fields of chemistry and microbiology is regarded as a positive starting point. The interdisciplinary orientation of synthetic biology requires a harmonised education approach for scientists and engineers alike.

▶ Similarly to genetic engineering as well as conventional breeding, synthetic biology is associated with risks relating to biological safety (biosafety) and possible abuse (biosecurity). There is still uncertainty as to whether the risks of synthetic biology require a different treatment and assessment compared to the risks of genetic research carried out so far. The initial assumption is that the existing rules and regulations are sufficient to avoid or minimise these risks. However, it is important that research on societal impact is carried out to help identify new risks at an early stage and thus prevent possible undesirable developments from the very beginning. With respect to biological safety, the risks of current research within the field of synthetic biology have been appropriately identified and regulated within a legal framework. Some of the approaches used in synthetic biology even contribute to increasing biosafety through the management of the viability of genetically modified organisms. One potential misuse of synthetic biology is the commercial acquisition of DNA sequences based on openly available genome data. However, Germany has already implemented legislation that limits this risk of misuse [Gentechnikgesetz (Genetic Engineering Act), Infektionsschutzgesetz (Infectious Diseases Protection Act), Kriegswaffenkontrollgesetz (War Weapons Control Act), Außenwirtschaftsgesetz (Foreign Trade and Payments Act)]. In addition to the legal regulations there are also voluntary undertakings applied within the scientific community and industry that relate to the management of toxins and pathogens together with the assessment of the trustworthiness of those who seek to purchase nucleic acid sequences. Researchers and manufacturers of synthetic nucleic acids have agreed that potential dangers arising from external orders for nucleic acid products must be determined and reduced by means of appropriate measures.

▶ The realisation that the boundaries between living matter and technically constructed matter are blurred in some areas has caused public concern that ethical boundaries are being crossed. Some have argued that the identity of living organisms is compromised when new types of life are created and that humankind is playing the role of Creator by tampering with nature. The counterargument is that influencing natural evolution is not at all ethically inadmissible and does not have to diminish our respect for life. Moreover, the exploitation of synthetic biology is associated with considerable potential benefits in fields such as medicine and environmental protection. From an ethical point of view, there is a need for an appropriate assessment and consideration of possible risks arising from synthetic biology. These and other issues must be discussed in a dialogue between all groups in society.

In summary, the present statement offers the following recommendations:

(1) Synthetic biology represents a logical further development of existing molecular biology methods and is associated with a large innovation potential from which both basic research and industrial applications can profit. Because the majority of the application-oriented projects are still at the design stage, basic research should be promoted and included to a greater extent in the planning of future scientific funding programmes.

(2) The success of synthetic biology will depend on the extent to which the various disciplines will be able to combine their intellectual capacities and to coordinate their infrastructures. Furthermore, young scientists should be given experience with this topic and its perspectives as part of their bachelor, master and graduate studies.

(3) With regard to the commercial exploitation of synthetic biology, it should also be taken into account that this utilisation depends not only on competent research, which must remain competitive within the international community, but also on the legal and societal framework conditions that play a decisive role in the success or failure of this new technology. It is expedient to carry out supporting research on the opportunities and risks at an early stage to ensure successful industrial utilisation of the new technology as well as its acceptance by society in general. This involves exploring what is socially acceptable in technical design in order to promote opportunities and reduce risks. The industrial application of the methods and products developed by synthetic biology should all be subject to the same protection under patent law as that applying to conventional recombinant gene products or gene fragments. It should be possible to protect minimal cells and protocells by copyright (preferably by patents) to provide an economic incentive for investments in the new techniques.

(4) With respect to biological safety (biosafety) and the risk of misuse (biosecurity), the existing legislation in Germany is sufficient for the present state of research. However, because of the dynamic and diverse developments, the following suggestions are made:

▶ the Zentrale Kommission für die Biologische Sicherheit (ZKBS, Central Committee on Biological Safety) should be commissioned to carry out scientific monitoring to support current developments in a professionally competent and critical manner and

▶ clearly defined criteria should be established to assess the risks relating to the release and handling under contained conditions of organisms used in synthetic biology for which there are no reference organisms in nature.

To reduce the risk of misuse, the following recommendations are proposed:

▶ establishment of a contact point with a standardised database that companies receiving questionable orders can consult, to assess DNA sequences and

▶ employees should be informed about possible risks of misuse of synthetic biology as part of their training according to the Gentechnik-Sicherheits-verordnung (Genetic Engineering Safety Ordinance).

▶ Should additional rules and regulations pertaining to risk assessment, monitoring and control of research and applications relating to synthetic biology become necessary in the course of development of the field, it is recommended that these should be drawn up on the basis of internationally recognised principles acting as a model for national rules and regulations.

(5) In cases for which proven methods of assessing the consequences of technology and risk analysis are not applicable or if the expected consequences are associated with large uncertainties, the precautionary principle must be applied. Furthermore, it is advisable that the self-regulation of science should be encouraged by establishing suitable interdisciplinary discussion platforms. Questions on the ethical evaluation of technically constructed life forms should be discussed openly at an early stage. This dialogue should include the exchange of perspectives and discussion of the various interpretations of living matter versus non-living matter. The aim of the discourse is the ethical evaluation of copied or de novo synthesised items that are integrated into naturally occurring organisms.

2 Introduction

A new research field known as synthetic biology has been emerging within an interdisciplinary context of biology, chemistry, physics, mathematics, engineering sciences, biotechnology and information technology.[1,2,3,4,5] Scientists from a variety of specialist fields are working together to design biological systems with new, defined properties. These systems use artificial means to obtain new biological components and novel living organisms that do not occur naturally in that form. Advanced methods of molecular biology, recombinant genetic engineering and chemical synthesis of biological building blocks are combined on the basis of scientific engineering principles. By this rational design, the assembly of synthetic and biological units should lead to new substances and systems, for example, novel polymeric molecules and tissues as well as entire cells and organisms.

But do these basic strategies upon which synthetic biology is based, as well as the resulting products, represent a new and revolutionary idea? As early as 1912, a publication by Stéphane Leduc contained the term "La Biologie Synthétique",[6] and in the same year, Jacques Loeb speculated that it should be possible to create artificial living systems.[7] After introduction of the term "synthetic biology" in the publications of Waclaw Szybalski,[8] today's understanding of the meaning of synthetic biology has been shaped in particular by Eric Kool's report, published by in 2000, on the incorporation of synthetic chemical components in biological systems.[9] Technological innovations in the synthesis of nucleic acids and DNA sequencing have certainly pushed the field of synthetic biology along a very fast and dynamic course. However, there is no clear line between synthetic biology and genetic engineering processes, which have been in use for over 30 years: for example in the synthesis of recombinant gene products.

1   Hartwell LH, Hopfield JJ, Leibler S, Murray AW; From molecular to modular cell biology. Nature, 1999, 402, C47–C52.

2   Benner SA, Sismour AM; Synthetic Biology. Nat. Rev. Genet., 2005, 6, 533–543.

3   Endy D; Foundations for engineering biology. Nature, 2005, 438, 449–453.

4   Andrianantoandro E, Basu S, Karig DK, Weiss R; Synthetic biology: new engineering rules for an emerging discipline. Mol. Syst. Biol., 2006, 2, 0028.

5   Heinemann M, Panke S; Synthetic Biology – putting engineering into biology. Bioinformatics, 2006, 22, 2790–2799.

6   Leduc S; La biologie synthétique. In: Études de biophysique. A. Poinat (ed.), Paris, 1912.

7   Loeb J; The mechanistic conception of life. In: Biological Essays. University of Chicago Press, Chicago, 1912.

8   Szybalski W; In vivo and in vitro Initiation of Transcription, 405. In: A. Kohn and A. Shatkay (Eds.), Control of Gene Expression, 23–24, and Discussion, 404–405 (Szybalski's concept of Synthetic Biology), 411–412, 415–417. New York: Plenum Press, 1974.

9   See, Rawls R; Synthetic Biology makes its debut. Chem. Eng. News, 2000, 78, 49–53.

The potential of synthetic biology is very diverse. This field makes a considerable contribution to acquiring basic research knowledge in that it attempts, for example, to answer questions regarding requirements for the viability of cells. Furthermore, synthetic biology opens new avenues for biotech applications, such as the development of improved pharmaceuticals, vaccines and diagnostic agents that are tailored to individual patients, the provision of synthetic gene vectors for successful genetic therapy as well as the design of specific biosensors, biological fuel cells and cell factories for the production of novel biomaterials. Synthetic biology includes methods for large-scale production of biofuels, such as ethanol, methanol and hydrogen, and processes to eliminate environmentally hazardous substances. It aims to modify specifically an organism's characteristics to produce fundamentally new properties that allow it to perform designed activities.

Chapter 3 of this statement discusses the scientific background for selected areas of synthetic biology and highlights their importance for the acquisition of general scientific knowledge. Six groups of topics are discussed in detail:

► Chemical-enzymatic synthesis of nucleic acids up to the level of a complete genome. This is a tool with which gene sequences can be specifically optimised and modified. The resulting products can be used, for example, to produce DNA vaccines and for somatic gene therapy.
► The construction of cells with a minimal genome. This genetic platform, also referred to as a "chassis", carries the minimum level of essential information for the viability of a cell. Minimal cells provide information on the evolutionary adaptation of organisms to natural habitats.
► Synthesis of protocells. Their construction plan follows either biological or physical principles. Protocells can be regarded as models of living cells.
► Production of biomolecules in previously unattainable quantities. Genetic engineering of complete metabolic reaction pathways using the modular assembly principle ("BioBricks") can provide a route to novel substances or production processes.
► Design of regulatory circuits. These are equipped with sensitive sensory functions and provide a means of network control for cellular or industrial processes.
► Use of modified cellular machines as part of a so-called "orthogonal system". This approach allows, for example, the design and production of polymeric compounds from chemical building blocks.

The extensive, and in some respects, still visionary spectrum of synthetic biology also raises a number of questions that are addressed in Chapter 4 of this statement:

► What are the economic benefits of synthetic biology and to what extent can society benefit from the new developments?
► Is there a risk of monopolies forming in this area of research?
► Does synthetic biology represent a particular risk potential that requires additional safety precautions or are the existing legal frameworks and the responsible monitoring bodies adequate for synthetic biology?
► Which ethical considerations are associated with synthetic biology, particularly projects that aim to produce artificial cells or to release novel organisms?

3   Selected Research Fields

3.1 Chemical Synthesis of Genes and Genomes

One of the key advances in the field of synthetic biology is that it is now possible to synthesise DNA with any sequence and in almost any length without a template, thus enabling the *de novo* synthesis of genes and even whole genomes. This means that new biological functions can be designed and further used for research and application purposes. Information provided by novel high throughput sequencing technologies is invaluable in this work.[10,11]

In conventional oligonucleotide synthesis, short-chain single-strand DNA molecules (~5 to ~50 nucleotides) are automatically synthesised as specific sequences. Gene synthesis then joins several oligonucleotides using sequential polymerase chain reactions, chip-based methods or by solid-phase assembly and plasmid cloning to produce long-chain synthetic DNA sequences.[12] This methodology can be used to produce several kilobases (kb) of genetic information according to the sequence specified by the genetic engineer. The upper limit is synthesis of the genome, which involves constructing *de novo* the full complement of genetic information of a virus, a bacterium, or, in future, even the minimal eukaryotic genome (see Section 3.2). Spectacular recent examples include the total synthesis of the poliomyelitis virus genome (~ 7.5 kb)[13] and the very much larger *Mycoplasma* genome (~ 583 kb).[14]

The new methods of synthesising large defined DNA fragments will decisively influence all research in the life sciences. Long-chain DNA sequences will become commercially available at high quality for every laboratory and for nearly every application. In the long term, this will save costs and reduce the time required to prepare genetic constructs.

---

10   Hall N; Advanced sequencing technologies and their wider impact in microbiology. J. Exp. Biol., 2007, 210, 1518–1525.

11   Church GM; Genomes for all. Sci. Am., 2006, 294, 46–54.

12   Tian J, Gong H, Sheng N, Zhou X, Gulari E, Gao X, Church G; Accurate multiplex gene synthesis from programmable DNA microchips. Nature, 2004, 432, 1050–1054.

13   Cello J, Aniko VP, Wimmer E; Chemical synthesis of poliovirus cDNA: Generation of infectious virus in the absence of natural template. Science, 2002, 297, 1016–1018.

14   Gibson DG, Benders GA, Andrews-Pfannkoch C, Denisova EA, Baden-Tillson H, Zaveri J, Stockwell TB, Brownley A, Thomas DW, Algire MA, Merryman C, Young L, Noskov VN, Glass JI, Venter JC, Hutchison CA 3rd, Smith HO; Complete chemical synthesis, assembly, and cloning of a Mycoplasma genitalium genome. Science, 2008, 319, 1215–1220.

The possibility of creating genetic information from potentially very pathogenic viruses by DNA synthesis, however, also incurs risks associated with misuse. In this respect, the successful ordering of DNA sequences suitable for making biological weapons by a British newspaper reporter caused alarm.[15] Therefore, suppliers of synthetic DNA should be subject to particular legal restraints, the scope of which is currently under debate. Leading commercial suppliers of synthetic DNA are trying to avoid potentially dangerous situations by means of self-regulatory codes of conduct (see Section 4.3).

The basic technological principles of gene synthesis were established more than 20 years ago. Technical advances have increased the productivity and quality of the processes with a continuous decrease in costs. Several dozen companies around the world currently offer commercial DNA sequences. These include flagship companies in Europe (Germany) and the USA. Whereas short DNA fragments of 0.1 to 1 kb can be delivered within a few days, the synthesis of a relatively large genome (for example a hypothetical minimal genome of ~110 kb)[16] with all the necessary quality controls currently needs up to one year. In comparison: the genome of the bacterium *Escherichia coli* K-12 comprises ~4.6 Mb and the human genome ~3000 Mb.

The chemical synthesis of DNA also allows the development of novel, sequence-optimised DNA libraries or the assembly of recombinant gene sequences that combine several artificially joined functional domains. Thus DNA synthesis can be used to prepare codon-optimised variants of human cDNA that, while conserving their natural amino acid sequence, have better expression properties after gene transfer into human or non-human cells.

Current applications of synthetic DNA in the field of pharmaceutical development relate to DNA vaccines and somatic gene therapy.

In the first example, DNA vaccines are used like conventional vaccines and induce the production of antigens by the body's own protein synthesis machinery. The resulting antigens provoke an immune response. For the vaccine manufacturer, this would eliminate the need for production and purification of the antigens on a large scale, thus resulting in greater flexibility in the selection of antigenic proteins. For example, the synthesis of a codon-optimised variant of genome sections of the human immunodeficiency virus type 1 (HIV-1) enabled the design of a complex DNA vaccine against HIV-1 that can present multiple antigens.[17] Its potential suitability to prevent an HIV-1 infection still has to be evaluated in comprehensive clinical studies. Moreover, the costs of large-scale production of DNA vaccines via chemical synthesis are still much too high.

In the second example, somatic gene therapy, the intention is to exploit the transfer of recombinant DNA in body cells to treat disease. Numerous appli-

---

15 Randerson J; Revealed: the lax laws that could allow assembly of deadly virus DNA. The Guardian, 14 June 2006; www.guardian.co.uk/world/2006/jun/14/terrorism.topstories3

16 Forster AS, Church GM; Towards synthesis of a minimal cell. Mol. Syst. Biol., 2006, 2, 45.

17 Bojak A, Wild J, Deml L, Wagner R. Impact of codon usage modification on T cell immunogenicity and longevity of HIV-1 gag-specific DNA vaccines. Intervirology, 2002, 45, 275–286.

cations for indications such as cancer and inflammatory, degenerative or monogenic disorders are at the preclinical development or clinical trial stages. In addition, novel, *in silico*-designed amino acid sequences are also more readily accessible by synthesis of extended gene sections and such "designer proteins" may exhibit antiviral activity. As with all applications in somatic gene therapy, the biological properties and possible toxicological or immunological reactions must be evaluated in comprehensive preclinical studies before they can be used in humans.

3.2 Development of Minimal Cells – Cells Reduced to Essential Vital Functions

One of the goals of synthetic biology is the development of so-called minimal cells that contain only those components that are absolutely essential for life. Minimal cells are defined by their minimal genome. A minimal genome contains only genes that are required for the survival of the respective organism under defined conditions. By generating minimal cells, it is possible not only to find out which genes of a living cell are essential under which conditions, but also to build a platform ("chassis") for new functions.

Comprehensive genome sequencing projects have shown that bacterial genomes have an extremely variable size. The first bacterial genomes sequenced – in the USA in 1995 – were the *Haemophilus influenzae* genome with 1.83 Mb[18] and the *Mycoplasma genitalium* genome with 0.58 Mb[19]. One of the pioneers of bacterial genome research is a German group who sequenced the *Mycoplasma pneumoniae* genome with 0.82 Mb.[20] Considerably smaller bacterial genomes have also been sequenced: the *Nanoarchaeum equitans* genome[21] with 0.49 Mb and the *Buchnera aphidicola* genome[22] with 0.42 Mb. The smallest bacterial ge-

18   Fleischmann RD, Adams MD, White O, Clayton, RA, Kirkness, EF, Kerlavage AR, Bult CJ, Tomb JF, Dougherty BA, Merrick JM et al.; Whole-genome random sequencing and assembly of Haemophilus influenzae Rd. Science, 1995, 269, 496–512.

19   Fraser CM, Gocayne JD, White O, Adams MD, Clayton RA, Fleischmann RD, Bult CJ, Kerlavage AR, Sutton G, Kelley JM, Fritchman RD, Weidman JF, Small KV, Sandusky M, Fuhrmann J, Nguyen D, Utterback TR, Saudek DM, Phillips CA, Merrick JM, Tomb JF, Dougherty BA, Bott KF, Hu PC, Lucier TS, Peterson SN, Smith HO, Hutchison CA 3rd, Venter JC; The minimal gene complement of Mycoplasma genitalium. Science, 1995, 270, 397–403.

20   Himmelreich R, Hilbert H, Plagens H, Pirkl E Li BC, Herrmann R; Complete Sequence analysis of the genome of the bacterium Mycoplasma pneumoniae. Nucl. Acids Res., 1996, 24, 4420–4449.

21   Waters E, Hohn MJ, Ahel I, Graham DE, Adams MD, Barnstead M, Beeson KY, Bibbs L, Bolanos R, Keller M, Kretz K, Lin X, Mathur E, Ni J, Podar M, Richardson T, Sutton GG, Simon M, Soll D, Stetter KO, Short JM, Noordewier M; The genome of Nanoarchaeum equitans: insights into early archaeal evolution and derived parasitism. Proc. Natl. Acad. Sci. USA, 2003, 22, 12984–12988.

22   Pérez-Brocal V, Gil R, Ramos S, Lamelas A, Postigo M, Michelena JM, Silva FJ, Moya A, Latorre A; A small microbial genome: the end of a long symbiotic relationship? Science, 2006, 314, 312–313.

nome known at present is that of the endosymbiont *Carsonella ruddii*[23], with a size of only ~0.16 Mb. The small genome size of all these bacteria is due to close adaptation to their specific host. However, such a lifestyle also means that these bacteria are difficult to handle experimentally, which is a major disadvantage in elucidating essential vital functions.

Minimal genomes can be developed using a *top-down* or a *bottom-up* approach. The *top-down* approach uses reduction of the existing genome, whereas *bottom-up* builds the minimal genome from individual DNA fragments.

In creating minimal cells, synthetic biology is primarily pursuing a scientific goal. The aim is to generate simplified cellular systems, and this also requires the acquisition of transcriptome, proteome and metabolome data, using mathematical modelling within the framework of systems biology. These cellular systems will help scientists to understand the systematic interplay of essential cell modules.

Of additional interest is an application-oriented goal to use minimal cells for various biotechnological production processes. Genetic components for desired metabolic functions can be inserted into the minimal genome of a cell that is to be used as a "chassis" and then optimised with respect to efficient production. But biological safety also plays a role in the development of production strains. When constructing a minimal genome, care should be taken to ensure that they do not carry any pathogenicity determinants. In addition it is also extremely important to appreciate that the ability of minimal cells to reproduce in the natural environment is very limited owing to the lack of all the genes that allow adaptation to complex and variable environmental conditions. Thus a minimal cell always has a reduced fitness compared to wild type cells and is thus particularly suitable, from the point of view of safety, for use in bioengineering processes and for deliberate release.

The *top-down* approach to creating minimal genomes has already been tested with several microorganisms: the gram-negative bacterium *Escherichia coli (E. coli)*[24], the gram-positive bacteria *Bacillus subtilis*[25] and *Corynebacterium glutamicum*[26] as well as the yeast *Saccharomyces cerevisiae*[27]. The genome is usually reduced by removing non-essential genes and intergene regions. These include,

23   Nakabachi A, Yamashita A, Toh H, Ishikawa H, Dunbar HE, Moran NA, Hattori M; The 160-kilobase genome of the bacterial endosymbiont Carsonella. Science, 2006, 314, 267.

24   Pósfai G, Plunkett G, Feher T, Frisch D, Keil GM, Umenhoffer K, Kolisnychenko V, Stahl B, Sharma SS, de Arruda M, Burland V, Harcum SW, Blattner FR; Emergent properties of reduced-genome Escherichia coli. Science, 2006, 312, 1044–1046.

25   Morimoto T, Kadoya R, Endo K, Tohata M, Sawada K, Liu S, Ozawa T, Kodama T, Kakeshita H, Kageyama Y, Manabe K, Kanaya S, Ara K, Ozaki K, Ogasawara N; Enhanced recombinant protein productivity by genome reduction in Bacillus subtilis. DNA Res., 2008, 15, 73–81.

26   Suzuki N, Nonaka H, Tsuge Y, Inui M, Yukawa H; New multiple-deletion method for the Corynebacterium glutamicum genome, using a mutant lox sequence. Appl. Env. Micr., 2005, 71, 8472–8480.

27   Murakami K, Tao E, Ito Y, Sugiyama M, Kaneko Y, Harashima S, Sumiya T, Nakamura A, Nishizawa M; Large scale deletions in the Saccharomyces cerevisiae genome create strains with altered regulation of carbon metabolism. Appl. Micr. Biotechnol. 2007, 75, 589–597.

for example, gene regions that allow utilisation of different food sources or elements that code responses to stress situations. Such non-essential gene regions can be identified using various techniques. Mutation analysis has proved very successful for this: transposons, for example, are used to mark the location of mutations. The annotated genome sequence is of decisive importance in switching off specific gene regions by means of deletion. An interesting side-effect results from the systematic deletion of insertion elements and transposons because this can be used to increase genome stability, which is important for technical applications. The *top-down* strategy to reduce the bacterial genome has been used extensively, notably with *E. coli*. The *E. coli* K-12 genome has been successfully reduced from 4.6 Mb to 3.7 Mb without loss of vitality.[28]

The *bottom-up* approach to creating minimal genomes starts with the conceptual design of a total sequence of a minimal genome. After complete chemical synthesis, it is transplanted into a cell envelope to enable cellular life. Such a *bottom-up* strategy is, without a doubt, a key element of synthetic biology but even the development of total sequences of minimal genomes on the drawing board requires tremendous knowledge of the interplay between individual cell modules. This knowledge is obtained using a wide variety of systems biology methods. Other key steps of the *bottom-up* approach have already been tested. For example, the group headed by Craig Venter successfully constructed the complete chemical synthesis of the 0.583 Mb genome of *Mycoplasma genitalium*.[29] This achievement can be regarded as a scientific breakthrough for this genome contains 5 to 7 kb sized DNA fragments that were assembled both *in vitro* and *in vivo*. Moreover, it had already been shown that a complete microbial genome can be transplanted into a cell envelope. This was achieved with the *Mycoplasma mycoides* genome that proved to be viable after transplantation into a *Mycoplasma capricolum* cell envelope.[30] Thus the first steps of the *bottom-up* approach to create synthetic minimal cells with minimal genomes have already been accomplished.

The interesting question now arises as to what size the minimal genome must have in order to carry out certain vital functions in the relevant organism. A satisfactory answer to this question can only be found if both the *top-down* and the *bottom-up* approach are followed in a combined strategy using selected organisms.

28 Pósfai G, Plunkett G, Feher T, Frisch D, Keil GM, Umenhoffer K, Kolisnychenko V, Stahl B, Sharma SS, de Arruda M, Burland V, Harcum SW, Blattner FR; Emergent properties of reduced-genome Escherichia coli. Science, 2006, 312, 1044–1046.

29 Gibson DG, Benders GA, Andrews-Pfannkoch C, Denisova EA, Baden-Tillson H, Zaveri J, Stockwell TB, Brownley A, Thomas DW, Algire MA, Merryman C, Young L, Noskov VN, Glass JI, Venter JC, Hutchison CA 3rd, Smith HO; Complete chemical synthesis, assembly, and cloning of a Mycoplasma genitalium genome. Science, 2008, 319, 1215–1220.

30 Lartigue C, Glass JI, Alperovich N, Pieper R, Parmar PP, Hutchison CA 3rd, Smith HO, Venter JC; Genome transplantation in bacteria: changing one species to another. Science, 2007, 317, 632–638.

3.3 Generation of Protocells – Artificial Systems with Properties of Living Cells

In contrast to minimal cells, protocells are not living cells, they are artificial units. They are self-replicating nanosystems constructed in the laboratory and exhibit many properties of living cells, for example, the existence of a mutatable information storage unit, a metabolic system and an enveloping membrane that contains the system but which is selectively open for the exchange of energy and materials with the surroundings. Protocells are regarded as a bridge between living and non-living matter.[31] The synthesis of protocells should help scientists understand the principles, modes of operation and origin of living cells. Thus the design of protocells represents a way of finding out which basic principles a living cell actually follows in order to function and develop. These questions also arise in the generation of minimal cells, which is why minimal cells are often grouped with protocells in the literature, although the terms are not synonymous.[32]

Bio-based protocells are constructed from the elementary building blocks of living cells (DNA, RNA, proteins, lipids). They can be regarded as possible precursors of living cells. A prominent example is the lipid membrane vesicles that enclose RNA replication systems, which are able to take up ribonucleotides. After fusing with the fatty acids in the surrounding medium, they are able to enlarge until they spontaneously divide into two "daughter cells".[33,34,35] It is also worth noting that cell-free expression systems (DNA → RNA → protein) can be encapsulated within the lipid membrane vesicle thus forming nanosystems which exhibit the characteristics of living cells.[36,37]

Artificial chemical-synthetic units with integrated complex electrical circuits have also been programmed as artificial cells that simulate the functions of living cells.[38]

31 Rasmussen S, Chen L, Deamer D, Krakauer DC, Packard NH, Stadler PF, Bedau MA; Transitions from nonliving to living matter. Science, 2004, 303, 963–965.

32 Rasmussen S, Bedau MA, Chen L, Deamer D, Krakauer DC, Packard NH and Stadler PF (eds.); Protocells. Bridging Nonliving and Living Matter. MIT Press, Cambridge, 2008.

33 Hanczyc MM, Fujikawa SM, Szostak JW; Experimental models of primitive cellular compartments: Encapsulation, growth and division. Science, 2003, 302, 618–622.

34 Chen IA, Roberts RW, Szostak JW; The emergence of competition between model protocells. Science, 2004, 305, 1474–1476.

35 Mansy SS, Schrum JP, Krishnamurthy M, Tobé S, Treco D, Szostak JW; Template-directed synthesis of a genetic polymer in a model protocell. Nature, 2008, 454, 122–125.

36 Ishikawa K, Sato K, Shima Y, Urabe I, Yomo T; Expression of a cascading genetic network within liposomes. FEBS Lett., 2004, 576, 387–390.

37 Noireaux V, Libchaber A; A vesicle bioreactor as a step toward an artificial cell assembly. Proc. Natl. Acad. Sci. U.S.A., 2004, 101, 17669–17674.

38 McCaskill, JS; Evolutionary microfluidic complementation towards artificial cells. in: Protocells. Bridging Nonliving and Living Matter. eds.: Rasmussen S, Bedau MA, Chen L, Deamer D, Krakauer DC, Packard NH and Stadler PF. MIT Press, Cambridge, 2008, 253–294.

In addition to their value in providing fundamental information, the development of protocells using different methods promises interesting new perspectives. For example, it may be possible synthetically to produce miniature factories for the production of pharmaceuticals and fine chemicals in protocells; however, this remains so far a visionary option.

According to current understanding, all organisms alive today originate from a primordial cell pool (progenotes) from which life on Earth developed about four billion years ago. Contemporary science is still a long way from being able to reproduce the evolution of life in a test tube and to construct living cells *de novo*. However, the synthesis of protocells begins to address the question as to where the boundaries between living and non-living matter lie and what life actually is. Some ethical guidelines have already been proposed in this regard.[39] Whether scientists exceed ethical boundaries by trying to synthesise living cells is still a matter of controversy. But we will have to answer this question should it indeed be possible to design living cells with new properties. This is discussed in more detail in Section 4.4.

3.4 Design of Tailored Metabolic Pathways

The design of tailored metabolic pathways *(metabolic engineering)* is often given as a typical example of synthetic biology. In the classical sense, this refers to the modification or supplementation of existing biosynthetic capacities either in organisms where some relevant metabolic steps already exist or in those where they are foreign. The targeted metabolic pathway is designed with controlling circuits and integration modules. The necessary DNA sequences are chemically synthesised, joined up (recombined) and then transferred into a suitable recipient organism.

The selective transfer of individual genes into foreign host organisms, such as the bacterium *Escherichia coli*, the yeast *Saccharomyces cerevisiae* or even human cells, has been common laboratory practice since the 1970s. This approach, which involves the transfer of DNA having several ten thousands of base pairs, is particularly attractive for the production of antibiotics and amino acids or in the development of transgenic plants.[40] The objective in these cases is to optimise the synthesis potential of a production strain. Thus *metabolic engineering* combines scientific interest with a commercial application.[41]

Additional routes have recently been opened for novel artificial biosynthesis processes that do not occur naturally. However, this approach is not really new technology, but rather represents an advancement of *metabolic engineering* as

39   Bedau MA, Parke EC, Tangen U, Hantsche-Tangen B; Ethical guidelines concerning artificial cells; www.istpace.org/Web_Final_Report/the_pace_report/Ethics_final/PACE_ethics.pdf

40   Rodriguez E, McDaniel R; Combinatorial biosynthesis of antimicrobials and other natural products. Curr. Opin. Microbiol., 2001, 4, 526–534.

41   Durot M, Bourguignon PY, Schachter V; Genome-scale models of bacterial metabolism: reconstruction and applications. FEMS Microbiol. Rev., 2009, 33, 164–190.

known since the mid-1980s. Although it was previously used for selective modification of individual genes or their regulators in a biosynthetic gene cluster comprising several genes, in 2003 this genetic engineering technique was used in *E. coli* to construct a complete biosynthetic pathway for producing isoprenoids. The bacterium was programmed so that it synthesised artemisinic acid, the precursor of the anti-malaria drug artemisinin.[42] This procedure involved recombining genes from the plant *Artemisia anna* and yeast as well as bacterial genes in *E. coli* together with the necessary bacterial control regions for regulated gene expression. Three years later, yeast was also successfully programmed to produce artemisinic acid.[43] The non-governmental organisation (NGO) One World Health, the biotech company Amrys, the Bill Gates Foundation and the pharmaceutical company Sanofi-Aventis are currently collaborating to implement this process. The objective of this work is to produce an anti-malaria drug that can be made available at low cost to patients in countries where malaria is endemic.

A further example is the synthesis of hydrocortisone from ethanol in yeast. In 2003, this process was achieved by a functional combination of 13 genes, eight of which were of human origin. Low-cost production is also the main objective in this case.[44] When compared to the total synthesis of hydrocortisone using the conventional method that involves more than 23 chemical and enzymatic reaction steps to obtain the end product, this process represents an important advance in the production method.[45]

In addition to the above-mentioned work in the field of pharmaceutical development, the construction of synthetic gene clusters and artificial biosynthetic routes are becoming increasingly important in the field of industrial or "white" biotechnology. Goals include the replacement of petrochemical-derived production processes by sustainable bioprocesses based on renewable resources. One example is the synthesis of a precursor to produce nylon.[46]

The transfer of a number of gene clusters that encode new natural substances into foreign host bacteria represents a remarkable advance. Furthermore, this technology can also be used to activate expression of "silent gene clusters". For example, functional expression of a gene cluster to produce a natural substance was achieved by transfer from the myxobacterium *Stigmatella* into *Pseudomonas*.

42   Martin VJ, Pitera DJ, Withers ST, Newman JD, Keasling JD; Engineering a mevalonate pathway in Escherichia coli for production of terpenoids. Nat. Biotechnol. 2003, 21, 796–802.

43   Ro DK, Paradise EM, Ouellet M, Fisher KJ, Newman KL, Ndungu JM, Ho KA, Eachus RA, Ham TS, Kirby J, Chang MC, Withers ST, Shiba Y, Sarpong R, Keasling JD; Production of the antimalarial drug precursor artemisinic acid in engineered yeast. Nature, 2006, 440, 940–943.

44   Szczebara FM, Chandelier C, Villeret C, Masurel A, Bourot S, Duport C, Blanchard S, Groisillier A, Testet E, Costaglioli P, Cauet G, Degryse E, Balbuena D, Winter J, Achstetter T, Spagnoli R, Pompon D, Dumas B; Total biosynthesis of hydrocortisone from a simple carbon source in yeast. Nat. Biotechnol., 2003, 21, 143–149.

45   Redaktion PROCESS; Hefezelle als Wirkstofffabrik. PROCESS, 22.02.2007; www.process. vogel.de/articles/58824/

46   Niu W, Draths KM, Frost JW; Benzene-free synthesis of adipic acid. Biotechnol. Prog., 2002, 18, 201–211.

This provides a means of specifically modifying the natural substance and obtaining considerably higher production yields.[47],[48] Subsequently, improved DNA transfer systems allow cloning of gene clusters of > 80 kb in *E. coli* and their expression in other host organisms such as *Streptomyces lividans*. This was recently demonstrated with the polyketide antibiotic meridamycin.[49]

There are, and will be, many other examples. What they have in common is that they are based on a detailed understanding of the biosynthetic pathways, a rational design and further development of the repertoire of experimental genetic engineering methods. In future, it will become standard practice to produce DNA more cheaply on the basis of available biosynthetic gene cluster sequences and their control elements, instead of using the time-consuming classical cloning route (see Section 3.1). Moreover, new techniques provide an opportunity optimally to adapt the genetic information to the production host. There are still many application potentials to be fully exploited.

The extent to which *metabolic engineering* can also be used in artificially constructed production hosts, such as minimal cells and protocells, will in turn depend on their production capabilities.

3.5 Construction of Complex Genetic Circuits

Since genetic circuits were described by Jacob and Monod[50] in the 1960s, molecular biologists have been interested in exploiting the wide range of possibilities for modifying cellular regulation processes and converting them into externally controllable genetic circuits.

DNA exerts its biological function via precise control of gene activity. Viruses, bacteria and eukaryotic cells use a wide range of complex regulatory mechanisms for this control that are encoded the nucleic acid level as regulatory motifs and which interact with cellular factors (RNAs or proteins). The gene activity can thus be closely harmonised to the metabolic and tissue-specific requirements of the cell at all levels of gene expression – from the formation of the primary transcript to the post-transcriptional modification (found in eukaryotes) and protein biosynthesis.

47   Wenzel SC, Gross F, Zhang Y, Fu J, Stewart AF, Müller R; Heterologous expression of a myxobacterial natural products assembly line in pseudomonads via red/ET recombineering. Chem. Biol., 2005, 12, 349–356.

48   Perlova O, Gerth K, Kuhlmann S, Zhang Y, Müller R; Novel expression hosts for complex secondary metabolite megasynthetases: Production of myxochromide in the thermopilic isolate Corallococcus macrosporus GT-2. Microb. Cell Fact., 2009, 8, 1–11.

49   Liu H, Jiang H, Haltli B, Kulowski K, Muszynska E, Feng X, Summers M, Young M, Graziani E, Koehn F, Carter GT, He M; Rapid cloning and heterologous expression of the meridamycin biosynthetic gene cluster using a versatile Escherichia coli-Streptomyces artificial chromosome vector, pSBAC (perpendicular). J. Nat. Prod., 2009, 72, 389–395.

50   Jacob F, Monod J; Genetic regulatory mechanisms in the synthesis of proteins. J. Mol. Biol., 1961, 3, 318–356.

Currently the most common artificial control system in biotechnology uses tetracycline-responsive promoters. These are based on the adaptation of a bacterial antibiotic-sensing system to control gene expression in cells. Tetracycline-responsive promoters have played a key role for many years in the functional analysis of genes and have potential for bioengineering product manufacture or therapeutic applications such as somatic gene therapy.[51]

There is no clear boundary between classical biotechnology and synthetic biology with respect to the development of artificial circuits. A number of other genetic circuits have been introduced in recent years into cells that regulate not only transcriptional control but also post-transcriptional mechanisms; the tetracycline-regulated system is still being subjected to comprehensive optimisation procedures.[52] If several of these circuits are combined, positive and negative feedback processes can be used to create complex cybernetic systems with differing characteristics. A key role is played by the so-called repressilator, which is an oscillating regulatory system based on the combination of three bacterial repressor proteins.[53] The construction of even more complex genetic circuits will benefit to an increasing extent from the development of functionally defined modules such as "BioBricks". But their interplay can only be predicted to a limited extent and must therefore be assessed empirically.[54,55]

In order to increase biological safety, as a basic principle, the regulation of organisms with artificial genetic circuits should depend on exogenically applied pharmaceuticals or other forms of chemical or physically defined induction.

3.6 Creation of Orthogonal Biosystems

Complexity plays a key role in the construction of novel biosystems: newly incorporated molecules or circuits interact with the existing system. The concept of orthogonal biosystems is used to integrate building blocks that should be as independent as possible from one another. One possible benefit is an improvement of biological safety.

Orthogonality refers here to the free combination of independent components. It is a construction principle from engineering sciences and plays a key role in information technology, for example. The strategy associated with orthogonality aims to modify subsystems without causing significant disturbances

51   Goosen M, Bujard H; Studying gene function in eukaryotes by conditional gene inactivation. Annu. Rev. Genet., 2002, 36, 153–173.

52   Greber D, Fussenegger M; Mammalian synthetic biology: engineering of sophisticated gene networks. J. Biotechnol., 2007, 130, 329–345.

53   Elowitz Mb, Leibler S; A synthetic oscillatory network of transcriptional regulators. Nature, 2000, 403, 335–338.

54   Stricker J, Cookson S, Bennett MR, Mather WH, Tsimring LS, Hasty J; A fast, robust and tunable synthetic gene oscillator. Nature, 2008, 456, 516–519.

55   Tigges M, Marquesz-Lago TT, Stelling J, Fussenegger M; A tunable synthetic mammalian oscillator. Nature, 2009, 457, 309–312.

in the other subsystems at the same time. The implementation of orthogonality in biological systems is regarded as a prerequisite for synthetic biology with respect to specific manipulations that go beyond a purely empirical approach and which are not subjected to the influence of cellular complexity.[56] In order to function independently, orthogonal subsystems should be as "invisible" as possible to the remainder of the cell, that is, they should exert only a minimal effect on the interactions of the natural (sub)systems.

One example is the engineering of the genetic code: proteins are generally composed of 20 different amino acids that determine their structure and function. There is of course no chemical or biological reason why amino acids other than the 20 "canonical" amino acids should not be used biologically as building blocks for proteins. Artificial amino acids can be inserted into selected positions of a protein for instance by modifying codons and correspondingly adapting the cellular translation machinery so that the genetic information is translated differently at the ribosome.

One way of specifically extending the genetic code to include an artificial amino acid is to use the least-used stop codon to insert this amino acid. This involves incorporating correspondingly modified transfer RNA (tRNA) and the loading enzyme into the cell. Ideally, this tRNA recognises only the stop codon and adds the artificial amino acid at this point during ribosomal protein synthesis without affecting the action of the already existing tRNAs.[57]

Another example of an orthogonal system is a modified ribosome that works with a quadruplet reading frame – that is, it translates four bases per codon instead of the usual three.[58] The aim is to establish two translation systems in a cell that work completely independently of each other: a "natural" translation system to synthesise normal cell proteins and an "orthogonal" system to synthesise polymers from amino acids that do not occur in nature. This method could be used to programme living cells to synthesise novel amino acid polymers. These new materials (including dental implants, replacement cartilage and bone) could then be used as therapeutic substances and for research purposes to elucidate structures and functions.

Orthogonal biosystems offer ways of increasing biological safety. For example, genes that are programmed for the synthesis of a specific gene product using an artificial genetic code can only be translated in organisms with the respective orthogonal translation system (see Section 4.3).

56   Panke S; Synthetic Biology – Engineering in Biotechnology. 2008, Swiss Academy of Technical Sciences (Ed.).

57   Budisa N, Weitze MD; Den Kode des Lebens erweitern. Spektrum der Wissenschaft, January 2009, 42–50.

58   Wang K, Neumann H, Peak-Chew SY, Chin JW; Evolved orthogonal ribosomes enhance the efficiency of synthetic genetic code expansion. Nat. Biotechnol., 2007, 25, 770–777.

4   Current Challenges

4.1 Economic Aspects

4.1.1 Market Potential

The economic prospects of synthetic biology can be assessed in terms of prospective commercial applications in the industrial and medical sectors as well as in licensing revenues and in the protection of intellectual property by means of patents. Even if synthetic biology is still in its infancy, attractive market potentials have already started to emerge. Economically interesting opportunities include increasing productivity by improving manufacturing processes, preparation of new products, shortening development times by standardising biological components and establishing new production concepts. Large market potentials for Germany as a producing country can be expected, particularly in the fields of white biotechnology, bioenergy and medicine. New production processes are emerging from the development of previously unknown synthetic routes as well as new ways of constructing production strains with improved properties. In addition, the provision of services is being developed in the fields of analysis and production of nucleic acids using technologies that are patent-protected.

The chemical industry, a traditional strength in Germany, is already profiting from the versatile processes of white biotechnology which demonstrate significant potential for creating new processes by means of synthetic biology. These processes may utilise new raw materials that not only conserve natural resources but also help to reduce waste. For example, the amino acid lysine, a nutritive additive, is currently produced with classical biotechnological methods at a rate of 700 000 tons per year, which corresponds to a market value of 1.4 billion Euro. In view of this high turnover, even the smallest optimisations in the bioengineering processes have high economic relevance. Thus *metabolic engineering* has considerable economic importance (see Section 3.4).

There is likely to be sequential change in the transition of a market economy and energy industry from that based on fossil fuels to renewable resources. From the economic point of view, the production of currently used starting materials based on renewable resources is initially expedient because this would allow continued use of existing production facilities. The medium-term goal should be to replace petrochemical resources by substances easily obtained by biological methods, which would entail a gradual change of the production processes and facilities.[59]

Synthetic biology promises new strategies for the production of biofuels. First-generation biofuels are based on plants that also serve as food. In view of the limited capacities of agricultural areas, this results in an undesirable competition with the cultivation of food crops. Processes to produce second-generation biofuels use the entire plant, particularly parts that are not used for food. These processes, such as the production of ethanol from agricultural wastes and plant residues, could be boosted by synthetic biology. The production of biohydrogen from water and solar energy using appropriately engineered microorganisms or biomimetic catalysts could also become a technically viable process in future. Research in these fields is being followed with great interest, and funded in some cases, by the large oil companies and the energy industry.

Diverse market potential is also envisaged for synthetic biology in medical diagnostics and prevention, development of pharmaceuticals and the use of alternative therapies. Potential applications in the fields of medicine, pharmaceutical development and the production of active substances have already been discussed in Sections 3.1 and 3.4.

Further support for interdisciplinary operating principles and early participation by scientific engineering capabiities is needed in order to develop new market potential into profitable applications and to accelerate the translation of basic knowledge into practice.

4.1.2 Patenting Issues

Genes and gene fragments that code a specific function can be patented. In Europe, this is regulated by EU Directive 98/44/EC[60] and its implementation in the European Patent Convention. This also applies to synthetic elements, including "BioBrick" parts. In 2007, the J. Craig Venter Institute attracted attention by applying for international and US patents: they claimed exclusive property rights to several essential genes of *Mycoplasma* and a synthetic organism (*Mycoplasma laboratorium*) that should be able to grow and self-replicate using these genes. The principle for securing property rights to genetically modified organisms (GMO) had already been established in 1980 by a decision of the Supreme Court of the United States in the case of Chakrabarty. The Court ruled that a GMO cannot be regarded as a product of nature and is therefore, in principle, patentable as long as further requirements (for example novelty value)

---

59   Biological modifications will increasingly drive the bioeconomy: biomaterials and bioenergy are estimated to make up one third of the industrial production capacity in Europe by 2030; see "En Route to the Knowledge-based Bio-Economy", Cologne Paper, May 2007.

60   Directive 98/44/EC of the European Parliament and of the Council of 6 July 1998 on the legal protection of biotechnological inventions; http://eur-lex.europa.eu/LexUriServ/LexUriServ.do?uri=OJ:L:1998:213:0013:0021:EN:PDF

are fulfilled.[61] In Europe, microbiological processes and products obtained with the help thereof qualify as patentable as a matter of principle (Art. 53b, EPC). Biological material that is isolated from its natural environment or is produced with the aid of a technical process is also patentable, even if it previously occurred in nature (Rule 27a, EPC).

Patenting of GMOs secures the holder– who has made an invention on the basis of significant research and by intellectual effort – a market lead by excluding others for a period of time from the commercial utilisation of the patented invention or by allowing such utilisation by means of a license. In addition, patents promote scientific development because the invention must be clearly and completely disclosed to an extent that it can be executed by a specialist; this provides the public access to knowledge that can be used as the basis for further developments and improvements. However, some have noted that there is risk of a monopoly on synthetic organisms, which could result in individual companies acquiring a dominating position.[62] This may be particularly critical if certain platform technologies become established as a standard. There are concerns expressed regarding limited access to socially important research materials and possible applications should the scope of the respective patents be too broad. A further issue is the potential development of so-called "patent thickets", as found in the electronics industry.[63] Because synthetic biology often requires a large number of "building blocks", the existence of numerous rights to these building blocks, which may be held by different entities, could hamper the development of new products.[64] To prevent this trend, some organisations, such as the non-profit BioBricks Foundation, offer freely accessible resources for synthetic biology to the public.[65] However, it is not always clear whether certain individual components of the available "BioBrick" parts have already been protected by a patent elsewhere.

Patents with a wide scope can indirectly hamper research inasmuch as commercial enterprises are less inclined to invest in research areas whose application-oriented implementation has already been extensively protected by patents. And the possibility of direct hindrance of research also cannot be denied. Actions for experimental purposes that relate to the subject of a patented invention are expressly excluded from the protection by the patent according to § 11 No. 2 Patentgesetz (Patent Act). The same applies to the utilisation of biological material for the purpose of culture, discovery and development of new types of plant (§ 11 No. 2a Patentgesetz (Patent Act)) as well as to studies and experiments and the resulting practical requirements that are necessary to obtain an

---

61   Diamond vs. Chakrabarty, 447 U.S. 303 (1980), US Supreme Court;
     http://caselaw.lp.findlaw.com/scripts/getcase.pl?navby=CASE&court=US&vol=447&page=303

62   See for example, ETC Group (Action group on Erosion, Technology and Concentration);
     Extreme Genetic Engineering: An Introduction to Synthetic Biology. 2007, 1–64.

63   Shapiro C; Navigating the Patent Thicket: Cross Licenses, Patent Pools, and Standard Setting.
     Innovation Policy and the Economy, 2000, 1, 119–150.

64   Henkel J, Maurer SM; The economics of synthetic biology. Mol. Syst. Biol., 2007, 3, 117.

65   http://bbf.openwetware.org/

authorisation under pharmaceutical law for placing on the market in the European Union or an approval under pharmaceutical law for the member states of the European Union or in non-EU countries (§ 11 No. 2b Patentgesetz (Patent Act)). The experimental use privilege is limited in that the experiments are only allowed if the patented object is used as the subject of the study and not simply as a means for their execution.

4.2 Research Funding and Education

Synthetic biology has been one of the focal points for research funding since about 2003. Since then, there has been a series of national research initiatives in a number of countries, including Great Britain, Denmark, The Netherlands, Switzerland, France and Germany. For example, the Cluster of Excellence "bioss" (Biological Signalling Studies) at the University of Freiburg that is part of the Excellence Initiative funded by the DFG. It combines methods of synthetic biology with studies on biological signal transmission.

European funding activities have also been specifically aimed at topics in synthetic biology. In the Sixth Framework Programme of the European Commission, 18 projects with a total budget of 24.7 million Euro were funded from 2007 to 2008 within the "NEST (New and Emerging Science and Technology) Pathfinder Initiative". These included not only projects aimed at the development of new products and methods but also projects for research communication (SynBio-Comm), for consideration of issues regarding biological safety and ethical aspects (SYNBIOSAFE) as well as strategic planning (TESSY – Towards a European Strategy for Synthetic Biology). It can be assumed that the NEST initiative, which finished in 2008/09, will be followed by new projects in the Seventh Framework Programme. Furthermore, the European Commission funded the integrated project "Programmable Artificial Cell Evolution" (PACE) from 2004 to 2008 and the project group published "Ethical guidelines concerning artificial cells", which gives an account of the current status of the discussions in this field.[66]

The European Science Foundation (ESF) has set up special funding programmes for synthetic biology projects, for example, a call for proposals for the EuroCore EuroSYNBIO (Synthetic Biology: Engineering Complex Biological Systems). The funds for this programme come from the respective funding organisations of the participating countries: from the DFG in Germany. In addition to these coordinated activities, there are also the individual funding projects of the DFG, which might deal with a topic related to synthetic biology.

Thus, there is a wide inventory of funding research in the field of synthetic biology which should be extended, if necessary. The success of all these funding activities is, however, decisively dependent on the following considerations:

---

66 Bedau MA, Parke EC, Tangen U, Hantsche-Tangen B; Ethical guidelines concerning artificial cells; www.istpace.org/Web_Final_Report/the_pace_report/Ethics_final/PACE_ethics.pdf

▶ interdisciplinary cooperation to generate synergies;

▶ the efficient use of available infrastructures and their expansion by concerted actions;

▶ a long-term view of basic research because many areas of synthetic biology are still at the fundamental research stage;

▶ integration of applied aspects into strategic planning at an early stage to facilitate and support industrial utilisation;

▶ guarantee transparency by means of communication that will foster public acceptance of this research field.

The success of synthetic biology will ultimately depend on the qualifications, fertile imagination and motivation of the next generation of scientists. This requirement necessitates inclusion of aspects of synthetic biology into the education curriculum of both scientists and engineers. The bachelor and master study programmes in Europe and an increasing number of graduate colleges and doctoral academies offer opportunities that have not yet been sufficiently exploited. Thus biologists should be given the opportunity at an early stage to deepen their basic knowledge of chemistry, physics and mathematics to improve their ability to think quantitatively. Equally, scientists who do not work in the life-science disciplines, as well as engineers, should be given insights into the physiology and biochemistry of living organisms and into techniques employed in molecular biology. This is indispensable for promoting efficient communication, a concerted approach and productivity.

This approach could also attract the interests of young scientists at an early stage in their education and encourage them to do interdisciplinary work. It could also promote their willingness to work in a team. One way of motivating them is the iGEM competition (international Genetically Engineered Machine Competition), which has been held since 2003 and where research groups from all over the world present their ideas related to synthetic biology for critical assessment.

Finally, graduates who have followed a challenging educational path should be offered attractive career prospects both in the academic and industrial fields.

4.3 Safety and Security Issues

Most of the avenues of research in synthetic biology listed in Section 3 employ biomolecular methods of genetic engineering. But the implementation of scientific engineering principles in synthetic biology introduces a new aspect that goes beyond genetic engineering.[67,68,69,70] In the opinion of some scientists, this

---

67 Forum Genforschung; Synthetic Biology. 2007, Platform of the Swiss Academy of Science.

68 Benner SA, Sismour AM; Synthetic Biology. Nat. Rev. Genet., 2005, 6, 533–543.

69 Heinemann M, Panke S; Synthetic Biology – putting engineering into biology. Bioinformatics, 2006, 22, 2790–2799.

70 Keasling JD; Synthetic biology for synthetic chemistry. ACS Chem. Biol., 2008, 3, 64–76

approach leads away from conventional analysis and modification and towards synthesising and constructing in synthetic biology.[71] From today's perspective, the aims of synthetic biology – to synthesise genomes *in vitro* and to create novel organisms do not yet mandate additional requirements for biological safety in laboratories or release (biosafety) and do not incur risks with respect to possible misuse (biosecurity) of this technology other than those arising from genetic engineering. Statutory regulation tailored to synthetic biology is thus currently not necessary.

However, owing to the fast rate of development, it is now recommended that work in the field of synthetic biology is monitored by the Zentrale Kommission für die Biologische Sicherheit (ZKBS, Central Committee on Biological Safety). It is also proposed that an official contact point is set up for companies working in the field of *in vitro* synthesis of nucleic acids to provide companies with information on potential risks of individual sequences. This would necessitate the establishment of a scientifically based and internationally harmonised database (see Section 4.3.3).

4.3.1 Biological Safety (Biosafety)

Biological systems are subject to the influence of diverse signals that are integrated via signalling components – similar to an electronic circuit system – into the network of the cell where they undergo evolutionary modifications. Unsuspected and new interactions might produce unexpected properties for artificial biological systems thus leading to incalculable risks if these systems are released either intentionally or unintentionally.[72,73,74,75]

Similar discussions on the complexity of biological systems and potential risks were held in the mid-1970s after DNA had been transferred for the first time across species boundaries from one organism to another.[76] The key risks arising from the production of genetically modified organisms were considered to be associated with their intentional and unintentional release with unpredictable consequences for the health of humans and animals as well as for the environment as a result of their interactions. These concerns have been and will be taken into account by means of a risk management concept that has been established for genetic engineering and which assumes that the risk hypothesised for genetic

71   van Est R, de Vriend H, Walhout B; Constructing Life. The World of Synthetic Biology. The Hague, Rathenau Institute. 2007, 1–16.

72   Bhutkar A; Synthetic biology: navigating the challenges ahead. J. Biolaw Bus., 2005, 8, 19–29.

73   Church G; Let us go forth and safely multiply. Nature, 2005, 438, 423.

74   Schmidt, M; SYNBIOSAFE – safety and ethical aspects of synthetic biology. 2007, Internet Communication.

75   Tucker JB, Zilinskas RA; The Promise and Perils of Synthetic Biology. The new Atlantis. Spring 2006, 25–45.

76   Berg P, Baltimore D, Boyer HW, Cohen SN, Davis RW, Hogness DS, Nathans D, Roblin R, Watson JD, Weissman S, Zinder ND; Letter: Potential biohazards of recombinant DNA molecules. Science, 1974, 185, 303.

engineering experiments actually exists (precautionary principle).[77] By working in risk-aware safety laboratories and by using procedures that progressively
transfer the GMO from the safety laboratory via, for example, a greenhouse
before releasing it to the environment, established approaches provide a means
for technical management of the risks assumed for the GMO. Vector-receiver
systems have been developed as biological safety measures with the help of biological safety research. These systems are not capable of reproducing themselves
outside a genetic engineering facility and have a limited life expectancy. They
also participate in horizontal gene transfer to a lesser extent than wild types of
organisms.[78] This risk management system for genetic engineering work and its
associated tools form the basis for assessing the risks of GMOs and for genetic
engineering work according to the Gentechnikgesetz (Genetic Engineering Act,
GenTG), which implements the EU directives 98/81/EC and 2001/18/EC.[79]

According to the GenTG, most of the work in the field of synthetic biology
described in Section 3 is genetic engineering work. New types of risks arising
from the large quantities of new recombinant nucleic acid sequences other than
those known from genetic engineering are not indicated on the basis of this
work; for many years, genetic engineering work has included the transfer of
nucleic acid segments of 50 kb to several 100 kb into cells via special vectors,
such as BACs or YACs.[80]

In those cases where organisms produced by synthetic biology for which no
characteristic reference organisms exist in nature are to be deliberately released,
the establishment of new evaluation systems (model ecosystems, e.g. micro-
and mesocosmos) for assessing risks should be considered before an approval
is issued for release into the environment. In this case, the GenTG can be used
as a basis for the characterisation of these organisms so that an appropriate risk
assessment can be carried out.

Some areas of synthetic biology do not necessarily fall within the scope of the
GenTG. For example, final evaluations are still outstanding for the *de novo* DNA
synthesis as a technique of modifying genetic material and for the assessment
of organisms produced by synthetic biology methods with a naturally occurring
sequence that was not assembled via recombination techniques. However, the
risks associated with these organisms can be assessed and controlled using the
tools of the GenTG. A future update of the GenTG should include an evaluation of the possible need for a more precise classification of organisms that
are not derived from natural organisms but which have been created *de novo*.

---

77 Berg P, Baltimore D, Brenner S, Roblin RO (III), Singer MF; Asilomar conference on recom-
    binant DNA molecules. Science, 1975, 188, 991–994.

78 Kruczek I, Buhk HJ; Risk evaluation. Methods Find Exp. Clin. Pharmacol., 1994, 16, 519–
    523.

79 Genetic Engineering Act in the version promulgated on 16 December 1993 (Federal Law Ga-
    zette I, p. 2066), last amended by Article 1 of the Law of 1 April 2008, Federal Law Gazette,
    499.

80 Burke DT, Carle GF, Olson MV; Cloning large segments of exogenous DNA into yeast by
    means of artificial chromosome vectors. Science, 1987, 263, 806–812.

The GenTG is currently not applicable to artificial cells, that is, those cells that are not able to reproduce themselves or transfer genetic material. However, these areas of synthetic biology are included in a risk assessment within the scope of the Chemikaliengesetz (Chemicals Act), the Arbeitsschutzgesetz (Occupational Safety Act) and – in the case of a pharmaceutical – the Arzneimittelgesetz (Medicinal Products Act) with the aim of protecting humans and the environment.[81,82,83] From the point of view of biological safety, neither cell-like systems nor subgenomic, replication-defective nucleic acids represent a potential hazard because they are not infectious or capable of reproduction, so that they cannot disseminate.

In conclusion, current work in the field of synthetic biology is integrated into a comprehensive assessment which is proportionate to the respective risk so that no new statutory regulations are regarded as necessary at the moment.

4.3.2 Synthetic Biology as a Means to Engineer Safety

The *de novo* synthesis of nucleic acids, described in Section 3.1 offers ways of increasing the safety with respect to intentional and unintentional release. Before an artificial nucleic acid is synthesised from chemical building blocks, the sequence is defined using a computer. Synthetically produced elements or organisms thus have a known nucleic acid sequence. The optimisation of *in vitro* synthesis of nucleic acids to produce increasingly longer sequences is a way of minimising rarely occurring cloning artefacts; it can also be used to avoid mobile genetic elements in synthetically produced genomes. Using *in vitro* DNA synthesis, non-natural nucleotides can also be used to produce components and organisms that are recognised only by specifically modified polymerases that do not occur naturally.

Independent of *in vitro* DNA synthesis, it is also possible to use non-natural amino acids that can only be incorporated into polypeptides by correspondingly modified ribosomes. Due to their dependence on artificial nutrients, these synthetic elements are not active in nature and/or the synthetically produced organisms are not capable of survival. The additional integration of synthetic circuits (see Section 3.5) or inactivation mechanisms into the genomes of artificially produced organisms coupled with the use of non-natural nutrients provides multiple protections. Thus, synthetic biology is based on biological safety concepts used in genetic engineering. But it also has the goal of developing the minimal cell, which can perform only limited functions in a defined environment and, so, reduces the hazard potential even further should it be released.

---

81   Chemicals Act in the version promulgated on 2 July 2008. Federal Law Gazette, 1146.

82   Medicinal Products Act in the version promulgated on 12 December 2005 (Federal Law Gazette, I 3394), last amended by Article 9, Paragraph 1 of the Law from 23.11.07. Federal Law Gazette, 2631.

83   Occupational Safety Act from 7 August 1996 (Federal Law Gazette, I, p. 1246), last amended by Article 15, Paragraph 89 of the Law from 5 February 2009. Federal Law Gazette, 160.

4.3.3 Protection against Misuse (Biosecurity)

Wilful misuse of biological substances and organisms for terrorist purposes is a latent threat that has been discussed in many variations and with a number of different scenarios (for example, attack scenarios with smallpox, ebola, anthrax, ricin). Suitable measures must therefore be implemented to provide protection against misuse.

New technical methods for genome sequencing and the provision of genome sequences in public databases facilitate general access to genetic data, including those of pathogenic organisms and biological toxins. The increasingly easy access to genome data and, in particular, the possibility of ordering defined nucleic acid sequences directly via the internet from companies that synthesise DNA are being debated as a specific hazard potential of synthetic biology.[84,85,86,87] In this respect, it should also be taken into consideration that recent virus research has led to the synthesis of a number of genomes from highly pathogenic organisms, including that of the poliomyelitis virus. It is feared that terrorist organisations or states will be able to reconstruct pathogenic organisms or toxins that can be used for hostile actions or acts of war. And, by analogy with computer hackers and authors of computer viruses, a similarly threatening approach could be followed by misguided individuals who are able to gain access to individual synthetic elements or to the necessary starting materials and produce synthetic systems or even microorganisms in an uncontrolled environment.

Because a pathogen has numerous characteristic properties (for example pathogenicity, infectiousness, host specificity), it is assumed to be less likely that new and more infectious pathogens could be created synthetically, but rather that existing pathogens can be reconstructed or modified (see Section 3.1). Owing to the high technical and logistical requirements, the likelihood of individuals misusing these techniques is estimated to be low.

As for all *dual use* technologies, protection against misuse or "biosecurity" in the field of synthetic biology follows the goal of minimising the possibility of misuse by means of specific measures. Germany has various statutory provisions that already greatly limit the risk of abusing synthetic biology. The Gentechnikgesetz (Genetic Engineering Act) grants approval to set up and operate a genetic engineering facility dependent on the trustworthiness of the operator and the persons responsible for its management and supervision. Furthermore, there must be no conditions that violate the agreement on chemical and biological weapons and with the Kriegswaffenkontrollgesetz (War Weapons Control Act).[88] According to the War Weapons Control Act, it is prohibited to develop,

---

84 Bhutkar A; Synthetic biology: navigating the challenges ahead. J. Biolaw. Bus., 2005, 8, 19–29.

85 Schmidt M; SYNBIOSAFE – safety and ethical aspects of synthetic biology. 2007, Internet Communication.

86 Schmidt M; Diffusion of synthetic biology: a challenge to biosafety. Syst. Synth. Biol., 2008, 2, 1–6.

87 Tucker JB, Zilinskas RA; The promise and perils of synthetic biology. New Atlantis, 2006, 12, 25–45.

88 Genetic Engineering Act in the version promulgated on 16 December 1993 (Federal Law Gazette I, p. 2066), last amended by Article 1 of the Law of 1 April 2008, Federal Law Gazette, 499. 1-4-2008.

manufacture or trade biological or chemical weapons in Germany. In addition, the Federal Republic of Germany has undertaken not to manufacture biological weapons given in the War Weapons List, which includes genetically modified microorganisms or genetic elements that are derived from the pathogenic microorganisms given in this list.[89] According to the Außenwirtschaftsgesetz (Foreign Trade and Payments Act), the export of genetic elements and genetically modified organisms in non-EU states requires an approval from the Bundesamt für Wirtschaft und Ausfuhrkontrolle (BAFA, Federal Office of Economics and Export Control).[90] The shipping of large DNA fragments is subject to special controls by the trade supervisory authority, the BAFA, the HADEX and K lists, which specifically limit the shipping of genes or gene fragments that could be used to manufacture biological weapons.

These regulatory instruments are additionally supported by voluntary undertakings of research bodies and industry. Using the Code of Conduct[91], published in April 2008 and relating to work with highly pathogenic microorganisms and toxins, the Deutsche Forschungsgemeinschaft (German Research Foundation) is attempting to focus the attention of scientists, in particular, to possible misuse of the work in this field and to provide information on how to deal with this issue.

Companies belonging to the International Association Synthetic Biology (IASB)[92] or the International Consortium for Polynucleotide Synthesis (ICPS)[93] have incorporated in their operating principles a commitment to investigate not only the origins of their customers but also the toxins and sequences to be synthesised with respect to pathogenicity factors and to decline any irregular or suspect orders. Some of the companies are following a very conservative policy: after reviewing the commissioned synthesis they will also decline an order to exclude the possibility of endangering their own workforce. As additional measures, it would be helpful to optimise and standardise the screening methods that are used to assess the DNA sequences for possible pathogenicity factors or toxins. This would probably necessitate a scientifically based database resource to allow standardised evaluation of DNA sequences; however, this should not be limited to Germany or even to Europe. Nonetheless, to resolve uncertainty, companies producing synthetic nucleic acids need a national contact point that they can contact if they encounter suspect orders.

The increasingly easy access to DNA sequences will lead to the adoption of techniques used in molecular biology and genetics in other scientific disciplines, such as the engineering sciences, where there is little experience in dealing with biological agents. In future, the Gentechnik-Sicherheitsverordnung (Genetic Engineering Safety Ordinance) will apply to the project management in these fields.

---

89   War Weapons Control Act in the version promulgated on 22 November 1990 (Federal Law Gazette I, p. 2506), last amended by Article 24 of the Ordinance from 31 October 2006. Federal Law Gazette, 2407. 2006.

90   Chemicals Act in the version promulgated on 26 July 2006. Federal Law Gazette, 1386. 2006.

91   www.dfg.de/aktuelles_presse/reden_stellungnahmen/2008/download/codex_dualuse_0804.pdf

92   www.ia-sb.eu/

93   http://pgen.us/ICPS.htm

4.3.4 Supervisory Monitoring

The fast and diverse developments in synthetic biology make it very difficult to predict whether other regulations will be necessary in future. Continuous scientific monitoring and, if necessary, evaluation of biological security issues are required. The legislative body should commission the Zentrale Kommission für die Biologische Sicherheit (ZKBS, Central Committee on Biological Safety) with security-relevant scientific support of synthetic biology. This Committee, which has been anchored in the GenTG since 1978, has been advising the Federal Government and federal states on security issues relating to genetic engineering. The ZKBS includes not only scientists but also other social groups representing, for example, occupational health and safety, consumer protection and environmental protection. Over the last 30 years, the ZKBS has cooperated with the authorities of the federal states and the government who are responsible for granting approvals and monitoring genetic engineering work and facilities and has helped to develop a socially acceptable system of assessing risks for organisms with engineered genomes. This system is continuously updated with the latest scientific and technical developments. On the basis of its specialist knowledge and expertise, the ZKBS is able to monitor the security-related scientific literature on synthetic biology. Furthermore, the contact point proposed above could cooperate with the ZKBS if new risks are recognised, in order to devise starting points for the adjustment of existing rules and regulations to address the potential hazards associated with synthetic biology.

As in genetic engineering, should it be necessary to draw up regulations for monitoring and controlling research and applications of synthetic biology that apply both on a national scale and for the individual states, they should be formulated on the basis of internationally recognised principles.

4.4 Ethical Issues

The issues discussed, for example, in Section 4.3 regarding unintentional damage or wilful misuse with respect to synthetic biology are just as relevant to ethical assessment as the issues relating to justice implied in Section 4.1 concerning intellectual property rights, patents and rights of use. Such problems have arisen from other sectors of modern biomedical research and should be discussed and treated against this historical background.

Since wide areas of synthetic biology represent a further development of molecular biology and genetic engineering, many well established methods of technology assessment and risk assessment are applicable. However, cases for which there are no natural reference systems require new risk assessment criteria because the development of new synthetic organisms is incalculable since their nature is still insufficiently explored. The rules of the precautionary principle should be applied, particularly in cases of high complexity and uncertainty. These steps include applying the principle of containment (spatial or temporal) to applications, close monitoring of the consequences and a flexible problem-oriented adjustment of the regulations to practical situations. Scenarios should be drawn up to assess the consequences, taking account of unintentional damage to humans, agriculture and the environment. Many risks can be reduced by implementing specific mechanisms of synthetic biology, for example, where the produced entities are not expected to survive outside the laboratory or do not participate in evolution. As an ethical principle, the possible damage (risk) must be balanced against the possible benefits (opportunities).

Many bioethicists see genuinely new ethical questions arising from the claim of creating new life by means of synthetic biology. These questions concern not only fundamental and new aspects of our understanding of life as opposed to artefacts or machines, but also the threat to, and value of, living matter and thus also to the self-image of humankind.[94] However, this hypothesised novelty of the ethical issues is contested by other bioethicists who see no need for special "synthetic bioethics"[95] and regard these questions as facets of known issues and wish to treat them as such – known from debates on the production of transgenic plants and animals, cloning, creation of chimeras or cell reprogramming as well as assisted reproduction and genetic enhancement.

Nevertheless, it is beyond dispute that these questions have to be formulated by ethicists and then brought to public discussion, which should take place before the planned technical advances are made – a platform for structured discussions could be set up.

Some hypotheses and goals can be formulated for the future debates.

(1) It is neither the objective nor realistic that synthetic biology will be able to create new higher organisms by synthesis or by manipulation in the foreseeable future. On the contrary, the goal is the modification and *de novo* synthesis of microorganisms, individual cells and cell populations. Nonetheless, this limited objective already poses basic questions regarding the definition of life, and the longer-term, at least hypothetical, options should be kept in mind.

(2) Our everyday preconceived understanding of "life" is determined by a number of partly inconsistent cultural and traditional criteria (morphological schemes, our understanding of nature – possibly guided by religious beliefs, general scientific education). Furthermore, various scientific disciplines with their specific research approaches and objectives start from a different understanding of life. For example, if the commonly used scientific concept relating to the definition of life based on a sustained metabolism, the ability to undergo evolutionary modification and to reproduce are generally taken as three necessary criteria for life, then mules, which like many hybrids[96] are not capable of reproduction, would not fall under the definition of a living organism (and would therefore not be covered by animal welfare acts, for example) – an ob-

---

94   See, Boldt J, Müller O; Newtons of the leaves of grass. Nat. Biotechnol., 2008, 26, 387–389; Boldt J, Müller O, Maio G; Synthetische Biologie. Eine ethisch-philosophische Analyse. 2009, Bern: Chapt. 6.

95   For example: Parens E, Johnston J, Moses J; Ethics. Do we need "synthetic bioethics"? Science, 2008, 321, 1449.

96   The mule is a hybrid bred from a female horse and a male donkey.

viously unreasonable result. A properly managed debate with understandable and reliable communications that addresses the challenges of synthetic biology thus requires a problem-oriented and consensus definition of living matter with the clear delimitation with respect to non-living matter. Semantic problems arise because many representatives of synthetic biology use terms and metaphors (for example "living machines") that appear to blur the boundary between living and non-living matter.

(3) The description of entities should include conceptual differentiation – *before* any evaluation – with respect to their properties, their functionalities and development potentials, and the circumstances of their origin (through natural processes, by synthesis or by genetic engineering). This is the only way to do justice to the potential complexity of conceivable forms of living matter.

(4) Moral arguments in favour of producing artificial life are based on the anticipated benefits for medicine, agriculture, energy generation or the environment, thus making the exploitation of synthetic biology not only permissible but even advisable. Although synthetic biology is justified by the economic advantages and ultimately by the freedom of research, according to the general consensus it should, however, be subject to other basic rights such as the right to physical integrity.

(5) The following could be included in the *fundamental* ethical objections *against* applications of synthetic biology:

(a) they inadmissibly intervene in creation or in sacrosanct processes of nature (playing God),

(b) by producing novel living organisms they destroy the integrity of nature or damage the order of living things and species or

(c) we would no longer respect and protect life in an appropriate manner in the course of developing "manufacturability".[97]

The first two types of objections are strongly based on ideological or metaphysical premises that are certainly not held by everyone, even within religious communities.

(a) Arguments concerning inadmissible intervention in creation or in natural processes are based on the religious concept that God alone may create life. In this case it is not the possible products of such interventions that are being criticised, but the process by which they are produced. But if we accept that the world was created by one God, it does not follow that humankind is forbidden to create life by artificial means. If it is assumed that all or some living organisms have their own independent intrinsic value, this does not exclude an extension to life produced artificially. And finally, a plausible case cannot be made as to why other profound interventions in nature (such as medical treatments) should then be rated more positively than the innovations of synthetic biology as a matter of principle.

---

97   See, Boldt J, Müller O, Maio G; Synthetische Biologie. Eine ethisch-philosophische Analyse. 2009, Bern: Chapt. 6.

(b) Also not convincing in this context are arguments suggesting that there are ethical problems regarding the production of *novel* living organisms, that is, those not previously occurring in nature. Thus the concept of nature's integrity being destroyed by humankind alone is hardly commensurate with our fundamental experience of nature's self-destruction, natural aggression, the occurrence of endemics, serious illnesses, etc. Moreover, the idea of a fixed and sacrosanct order of living beings and species is already contradicted by natural biological phenomena relating to changes, intermixing or the extinction of species.

(c) The objection that applications of synthetic biology could have a negative impact on our basic understanding of life in general and on the worthiness of protecting human life in particular, certainly requires a thorough analysis on the basis of its threatening potential, but it does appear to be rather speculative at first sight.

(6) The debate on the self-regulation of science is also a matter of controversy, particularly with regard to synthetic biology. The responsible implementation of the freedom of research is taken seriously by the scientific community. At the "SyntheticBiology 2.0" conference in Berkeley in 2006, concepts for self-governance were discussed and presented to the public[98]. One of the main goals was to search for ways in which to find a balance between the free availability of data and how to prevent its misuse. However, in an open letter, 35 NGOs criticised this self-governance approach as inadequate and called for a wider, inclusive public dialogue.[99] The letter draws parallels to the "Asilomar Conference on Recombinant DNA" held in 1975 at which a group of 140 scientists called for self-governance on the basis of the precautionary principle with respect to managing recombinant DNA. The effect of this declaration left the oversight of genetic technology to a great extent in the hands of scientists for a long time. The approaches for self-regulation by the scientific community have not been continued since the above-mentioned concept was proposed.

(7) All these considerations relating to the innovative research field of synthetic biology should be discussed on a comprehensive, interdisciplinary and intercontextual basis. This requires supplementary ethical research at an early stage as well as critical reflection on the responsible implementation of the freedom of research in scientific fields. Furthermore, intensive efforts should be made to educate the public early on about the work in the laboratory and about the risks and benefits so as to provide a basis for ethical reflection.[100]

---

98   See, Schmidt M, Torgersen H, Ganguli-Mitra A, Kelle A, Deplazes A, Biller-Andorno N; "SYNBIOSAFE e-conference: online community discussion on the societal aspects of synthetic biology", in: Systems and Synthetic Biology (Online First Publication, 2008 Sep 18): 11 pages. Public declaration from the Second International Meeting on Synthetic Biology (May 20–22, 2006, Berkeley, CA); http://hdl.handle.net/1721.1/32982

99   NEWS RELEASE, 19th May 2006, Global Coalition Sounds the Alarm on Synthetic Biology, Demands Oversight and Societal Debate; www.etcgroup.org/en/materials/publications.html?pub_id=8

100 See also Schmidt M, Torgersen H, Ganguli-Mitra A, Kelle A, Deplazes A, Biller-Andorno N; "SYNBIOSAFE e-conference: online community discussion on the societal aspects of synthetic biology". In: Systems and Synthetic Biology (Online First Publication, 2008 Sep 18): 11 pages.

Appendix

A) Genesis of the Statement and Members of the Working Group

The three participating organisations (DFG, acatech and Leopoldina) organised a joint workshop that was held on 27 February 2009 in Berlin (programme: see Appendix B). The speakers and participants laid the foundation for this statement with their presentations and contributions to the discussion. This statement was drawn up at the conclusion of the workshop by the interdisciplinary working group "Synthetic Biology" headed by Prof. Dr. Bärbel Friedrich, Chairwoman of the Senate Commission on Genetic Research of the DFG. The members of the working group are listed below. The individual parts of the text were written in consultation with further experts. This statement was subjected to a review process and then adopted by the Executive Committees of the DFG, acatech and Leopoldina.

Members of the Working Group

Professor Dr. Christopher Baum Member of the Senate Commission on Genetic Research	Medizinische Hochschule Hannover Abteilung Experimentelle Hämatologie Carl-Neuberg-Straße 1 / OE 6960, K 11, Raum 1120 30625 Hannover (Germany)
Dr. Matthias Brigulla	Bundesamt für Verbraucherschutz Referat 402 Mauerstraße 39–42 10117 Berlin (Germany) (currently seconded to the Federal Ministry of Food, Agriculture and Consumer Protection)
Professor Dr. Bärbel Friedrich Member of the Senate Commission on Genetic Research Vice-President of Leopoldina Chairwoman of the Working Group	Humboldt-Universität zu Berlin Institut für Biologie Chausseestraße 117 10115 Berlin (Germany)
Professor Dr. Carl F. Gethmann Member of acatech Member of Leopoldina	Universität Duisburg-Essen Fachbereich Geisteswissenschaften Institut für Philosophie Universitätsstraße 12 45141 Essen (Germany)

Professor Dr. Jörg Hacker Vice-President of the DFG Member of the Senate Commission on Genetic Research Member of Leopoldina	Robert Koch-Institut (RKI) Nordufer 20 13353 Berlin (Germany)
Professor Dr. Klaus-Peter Koller Member of the Senate Commission on Genetic Research	Sanofi-Aventis Deutschland GmbH F&E, External Innovation Bldg. H 831 Industriepark Höchst 65926 Frankfurt (Germany)
Professor Dr. Bernd Müller-Röber Member of acatech	Universität Potsdam Mathematisch-Naturwissenschaftliche Fakultät Institut für Biochemie und Biologie Karl-Liebknecht-Straße 24–25 14476 Golm (Germany)
Professor Dr. Alfred Pühler Member of acatech Member of Leopoldina	Universität Bielefeld Centrum für Biotechnologie (CeBiTec) Universitätsstraße 27 33615 Bielefeld (Germany)
Professor Dr. Bettina Schöne-Seifert	Universitätsklinikum Münster Institut für Ethik, Geschichte und Theorie der Medizin Von-Esmarch-Straße 62 48149 Münster (Germany)
Professor Dr. Jochen Taupitz	Universität Mannheim Institut für Deutsches, Europäisches und Internationales Medizinrecht, Gesundheitsrecht und Bioethik der Universitäten Heidelberg und Mannheim Schloss / Postfach 68131 Mannheim (Germany)
Professor Dr. Rudolf Thauer Member of the Executive Committee of Leopoldina	Max-Planck-Institut für terrestrische Mikrobiologie Karl-von-Frisch-Straße 35043 Marburg (Germany)
Professor Dr. Angelika Vallbracht Member of the Senate Commission on Genetic Research	Universität Bremen Zentrum für Umweltforschung und nachhaltige Technologien (UFT) Abteilung Institut für Virologie Postfach 330440 28359 Bremen (Germany)

From the Head Offices

Dr. Ingrid Ohlert	Deutsche Forschungsgemeinschaft (German Research Foundation) Fachgruppe Lebenswissenschaften Kennedyallee 40 53175 Bonn (Germany)
Dr. Nikolai Raffler	Deutsche Forschungsgemeinschaft (German Research Foundation) Fachgruppe Lebenswissenschaften Kennedyallee 40 53175 Bonn (Germany)
Dr. Marc-Denis Weitze	acatech – Deutsche Akademie der Technikwissenschaften Projektzentrum Residenz München Hofgartenstraße 2 80539 München (Germany)

For their support in drawing up this statement, we thank Prof. Dr. Nediljko Budisa (Max-Planck-Institut für Biochemie, Planegg), Dr. Jürgen Eck (B.R.A.I.N. AG, Darmstadt), Dr. Margret Engelhard (Europäische Akademie zur Erforschung von Folgen wissenschaftlich-technischer Entwicklungen Bad Neuenahr-Ahrweiler GmbH), Prof. Dr. Jürgen Heesemann (Universität München), Prof. Dr. Hans-Dieter Klenk (Universität Marburg) and Prof. Dr. Petra Schwille (Technische Universität Dresden). We thank Dr. Robin Fears (EASAC London, UK) for his editorial contribution.

Special thanks go to the members of the Senate Commission on Genetic Research who initiated the drawing up of this statement within the DFG, for their useful comments and their support *(www.dfg.de/dfg_im_profil/struktur/gremien/senat/kommissionen_ausschuesse/senatskommission_grundsatzfragen_genforschung/index.html)*.

We are also grateful to the speakers and contributors to the „Synthetic Biology" workshop held on 27 February in Berlin (Germany) for their work during the preparatory phase.

B)   Workshop Programme

"Synthetic Biology" Workshop

Thursday, 26 February 2009 – Hotel NH Berlin Mitte (Leipziger Straße 106–111)

Arrival of participants	
19:30 – 21:00	Reception at the Hotel (Prof. Matthias Kleiner)

Friday, 27 February 2009 – Landesvertretung Sachsen-Anhalt (Luisenstraße 18)

09:00 – 09:20	Welcome (Prof. Matthias Kleiner, Prof. Reinhard Hüttl, Prof. Bärbel Friedrich)
Part I	**Moderation: Prof. Jörg Hacker**
09:20 – 09:55	Minimal Genomes (Prof. György Pósfai, Szeged, HU)
09:55 – 10:30	Protocells (Prof. John McCaskill, Bochum, DE)
10:30 – 11:05	Orthogonal Biosystems (Prof. Jason Chin, Cambridge, UK)
11:05 – 11:25	Coffee Break
11:25 – 12:00	Genetic Circuits (Prof. Martin Fussenegger, Zürich, CH)
12:00 – 13:30	Discussion (Prof. Alfred Pühler)
13:30 – 14:30	Lunch Break
Part II	**Moderation: Prof. Bärbel Friedrich**
14:30 – 15:05	Ethical Issues (Prof. Paul Martin, Nottingham, UK)
15:05 – 15:40	Socioeconomical Issues (Prof. Ralf Wagner, GeneArt, Regensburg, DE)
15:40 – 16:15	Legal Issues (Dr. Berthold Rutz, European Patent Office, München, DE)
16:15 – 16:45	Coffee Break
16:45 – 17:20	Biosafety and Biosecurity Issues (Dr. Markus Schmidt, Vienna, AT)
17:20 – 18:50	Discussion (Prof. Klaus-Peter Koller)
18:50 – 19:00	Closing Remarks (Prof. Rudolf Thauer)
19:00 – 21:00	Dinner – Restaurant "Habel Weinkultur" (Luisenstraße 19)

Saturday, 28 February 2009 – Hotel NH Berlin Mitte (Leipziger Straße 106–111)

Departure	

C)   Glossary

BAC: Bacterial Artificial Chromosome; vector used for → Cloning large sections of the genome in yeast cells, for example *Escherichia coli.*

BAFA: Bundesamt für Wirtschaft und Ausfuhrkontrolle (Federal Office of Economics and Export Control)

BioBrick: Characteristic genetic building block or genetic circuit element.

BMBF: Bundesministerium für Bildung und Forschung (Federal Ministry of Education and Research)

bp: Base pair

cDNA: *complementary DNA.* This is a type of → DNA that is usually synthesised from → mRNA by the reverse transcriptase enzyme.

Clone: → Cloning

Cloning: The process of copying and reproduction of identical individuals. This term is used for molecules, cells, tissues, plants (offshoots), animals and humans. Clones are genetically identical copies.

Codon: Designation for a sequence of three → Nucleobases (base triplet) that → mRNA encodes for a specific amino acid in the genetic code.

de novo: (lat.) from new (from scratch)

DNA: Deoxyribonucleic acid; chemical building block of the genetic material. The DNA contains the information required to synthesise all the proteins needed for the bodily functions.

EPC: European Patent Convention

Expression: → Gene expression is the conversion of the information stored in the DNA of a gene to produce cell structures and signalling molecules. These are often proteins. The expression of genes is a complex process comprising many different individual steps. In general, regulation of gene expression at various levels of the conversion process can lead from the gene to the particular characteristic.

Gene: Section of the DNA that is coded for a particular function, e.g. a protein. In addition to the coded areas (exons), the genes also have other regions such as introns (non-coded areas) and → Promoters (regulation elements).

Gene expression: Conversion of genetic information, usually in the form of proteins, to build cell structures and signalling molecules.

Genome: Inconsistently used term for the total → DNA of an individual or the genetic information in a cell (→ Gene).

Genetic engineering: Bioengineering methods and processes of biotechnology that allow specific manipulation of the genetic material (→ Genome) and thus of the biochemical control processes in living organisms or viral genomes.

GenTG: Gentechnikgesetz (Genetic Engineering Act)

Gene therapy: → Somatic gene therapy

Gene transfer: The procedure of introducing genes into cells.

GenTSV: Gentechnik-Sicherheitsverordnung (Genetic Engineering Safety Ordinance)

GMO: Genetically modified organism. Organism whose genetic material has been specifically modified by means of genetic engineering methods.

HADEX-Liste: An exclusion list naming customers (companies, institutions) that must not be have access to *dual use* goods.

IASB: International Association Synthetic Biology

ICPS: International Consortium for Polynucleotide Synthesis

in silico: (based on lat., *in silicio,* in silicon); processes performed on a computer.

in vitro: (lat.) within the glass (test tubes, in tissue cultures, etc.); this refers to operations outside the organism, as opposed to → *in vivo,* within a living organism.

in vivo: (lat.) within the living; processes occurring within a living organism.

Insertion: Insertion of DNA sequences into a genome.

kb: Kilobases = 1000 bases

K-Liste: An exclusion list naming countries that should not have access to *dual use* goods.

Mb: Megabases = 1 000 000 bases

Metabolome: Complete set of all metabolites

mRNA: (messenger RNA); designation for the → Transcript from a subsection of DNA belonging to one → Gene.

NGO: non-governmental organisation

Nucleic acid: Macromolecules made up of individual building blocks, the → Nucleotides. See also → DNA.

Nucleobase: → Nucleotide

Nucleotide: Basic building block of a → Nucleic acid (→ DNA and → RNA).

Oligonucleotide: (Greek *oligo*, few); oligomers made up of a few → Nucleotides (→ DNA or → RNA).

PatG: Patentgesetz (Patent Act)

Plasmid: Small, generally circular DNA molecule that is capable of replicating independently. Plasmids are found extrachromosomally in bacteria. They may contain several → Genes.

Promoter: (originally French *promoteur*, initiator); designation for a DNA sequence that allows regulated → Expression of a → Gene. The promoter sequence is an essential component of the gene.

Proteome: Complete set of → Proteins

Ribosome: (Greek Αραβινός, *arabinos*, grape and σωμα, *soma*, body); highly specialised complex consisting of proteins and RNA. It plays a central role in protein biosynthesis: the information stored in the → mRNA sequence is read off and used to synthesise proteins.

RNA: Ribonucleic acid; stores information in the form of nucleic acids. Plays a key role in the conversion of genetic information into proteins (→ Transcription).

rRNA: Ribosomal RNA

Somatic cells: Cells in the body whose genetic information cannot be inherited by subsequent generations. They make up the majority of cells in the human body. Only germ cells (eggs and sperm cells) can transfer genetic information to the next generation thus forming a so-called germ line (→ Somatic gene therapy).

Somatic gene therapy: Application of gene transfer to → Somatic cells with the objective of preventing or treating illnesses. In this case, genetic modifications are not passed on to the offspring.

Systems biology: Branch of biosciences that attempts to understand biological systems and processes quantitatively in their entirety.

Transcript: → Transcription

Transcription: (lat. *trans,* across, beyond, on the opposite side; *scribere,* write); in biology, transcription refers to the first step of protein biosynthesis that leads to the formation of → mRNA; it also includes the synthesis of → tRNA and → rRNA. In the transcription process, a → Gene is read off and copied as an mRNA molecule, in other words, a specific section of DNA acts as the template for the synthesis of a new RNA strand. This process involves the transcription of the nucleic bases of DNA (T, A, G, C) into the nucleic bases of RNA (U, A, G, C).

Transcriptome: Complete set of → Transcripts

Transgenic: This is usually a genetically modified organism (→ GMO), which contains additional → Genes from other species in its → Genome.

Transposon: Section of a gene that can move to a different position within the → Genome (= transposition).

tRNA: transfer RNA

Vaccine: A biological or genetically engineered antigen, usually consisting of protein or genetic fragments of weakened or dead pathogens. Inoculation with the vaccine causes specific activation of the immune system against a particular pathogen or group of pathogens.

YAC: Yeast Artificial Chromosome; vector used for → Cloning large sections of the genome in yeast cells.

ZKBS: Zentrale Kommission für die Biologische Sicherheit (ZKBS, Central Committee on Biological Safety)